THE
DEVIL MADE
CRYPTO

LAFLIN HILL
PRESS

Contact information for Laflin Hill Press– inquires@laflinhillpress.com

ISBN: 979-8-9872246-0-1 (paperback)
ISBN: 979-8-9872246-1-8 (ebook)
ISBN: 979-8-9872246-2-5 (hardcover)

Ordering Information:
Special discounts are available on quantity purchases by corporations, associations, and others. For details, contact www.lerntt.com

THE
DEVIL MADE
CRYPTO

SEEKING THE TRUTH ABOUT WHY PEOPLE THROW SHADE **ON CRYPTOCURRENCY**

ANDREW JENKINS

SPECIAL THANKS

Thank you, God, for giving me the strength to write this book.

DISCLAIMER

This is not a technical book, so don't worry about having to decipher a bunch of technical terms and crypto jargon! Easy read!

BITCOIN WAS BORN

On October 31, 2008, a stranger using the screen name Satoshi Nakamoto posted a whitepaper titled "Bitcoin: A Peer-to-Peer Electronic Cash System" to the Cryptography mailing list, and introduced Bitcoin as the solution to a fundamental problem with the banking system—the double-spending problem.[1] In the whitepaper, Satoshi deduced that creating a digital currency that ran on a decentralized, peer-to-peer public network eliminated the need for buyers and sellers to trust third-party financial intermediaries with their transactions. People were captivated by the whitepaper. Two months later, on January 3, 2009, Bitcoin was born.[2]

Then on April 26, 2011, Satoshi sent his final email correspondence, never to be heard from again.[3]

CONTENTS

I LIED TO YOU

O kay, it's probably not the best way to kick off our relationship, but if you give me a chance to explain, you'll see where I'm coming from.

The title of this book is *The Devil Made Crypto*, but I must confess, that is a lie. The truth is, the devil did NOT make cryptocurrency. But would you believe that some individuals, governments, financial institutions, and whale-sized investors treat cryptocurrency as if it were the evilest invention ever made? This is true!

I know most of you have seen or at least heard some of the shade that politicians, investors, and even some of your disgruntled family members have thrown on cryptocurrency:

- *Only criminals use crypto...*

- *Crypto is the number one killer of the environment...*

- *Crypto is nothing but a scam...*

- *Crypto will never have a place in the world...*

- *Crypto is BAD!*

I'll have you know that there is more to the story. Cryptocurrency is doing more good in the world than naysayers would have you believe.

In this book I will shed light on how people's fears, biases, and obsession to control threaten the future of decentralized finance and innovation, and how they rob disadvantaged societies from obtaining the financial freedoms they desperately need.

We will be looking at this problem from the perspectives of individuals, institutions, and crypto protocols:

Individuals – some people turn against the crypto industry because they have unresolved biases, fears, and egos.

Institutions – governments, governmental organizations, and financial institutions view cryptocurrency as a threat to their ability to control the financial system, so they create false narratives, policies, and laws to discredit cryptocurrency and stifle its evolution.

Crypto protocols – unfortunately, some crypto protocols function in a way that cause problems for people in the crypto community. The problems open the door for bad actors to exploit vulnerable networks and give the crypto industry a bad name.

I will also take you through the exciting history of cryptocurrency from the Cypherpunks movement and David Chaum to Bitcoin creator Satoshi Nakamoto. You will learn about some of the fantastic innovations that states and countries have made to revolutionize their economies and solve problems in the crypto industry.

Lastly, I will provide steps that the crypto-community can take to minimize opposition toward cryptocurrency and gain support from the broader market.

CHAPTER ONE

DEMONS

S o, if the devil made crypto, then who are the demons that laid the groundwork for crypto to exist?

Think about the first smartphone you purchased. You probably didn't spend six painstaking weeks perusing every scholarly article and peer-reviewed journal about the designers' workforce capabilities, technical inspirations, obstacles, or the oppositions they overcame. You saw a handheld computer with voice, text, photo, video, and internet capabilities all in one and bought it as quickly as humanly possible, right?

Would your appreciation for your phone change if someone told you about the years of painstaking research it took to create every little component for the phone to work? Probably so.

I would want you to have that same appreciation for crypto. However, to appreciate how bona fide cryptocurrency is, you deserve a better explanation than the clichéd *"It puts the power in the hands of the people"* line that most people would give you. Of course this is true, and it explains why many communities around the world long for it, but there is an entire history behind cryptocurrency that oftentimes goes unexplored. I want to provide you with some of it.

Like many other crypto-traders, my initial understanding of cryptocurrency was that the whole thing—blockchain, anonymous transactions, trustless verification, electronic privacy systems, etcetera—started in 2009 with the creation of Bitcoin. But I later learned, as I prepared to write this chapter, that Bitcoin's story—and all its associated technologies—began decades before 2009 with the advent of robust encryption software and the internet.

For decades, inventors, computer scientists, mathematicians, and software developers bolstered cryptography to enhance privacy individualism—which ultimately paved the way for Bitcoin and alternative cryptocurrencies.

I'll be the first to admit that origin stories can sometimes be boring (unless you're watching a Marvel movie (Disney, not Sony)). This is especially true when it comes to the complexity of technology. But the people who pioneered crypto technology were not your typical computer scientists. These guys were activists, revolutionaries, and anarchists who called the public to resist electronic government surveillance and hold dear to individual rights. They were the *demons* that challenged government control over society. They were the *Cypherpunks.*

The Cypherpunks Movement

The Cypherpunks movement of the nineties is still one of modern history's most significant techno-political movements. They represented something radical, a twentieth-century embodiment of John Locke that adopted the philosophy that protecting individual freedom and civil liberty takes precedence over authoritative rule.[4]

The Cypherpunks were not, as I initially assumed, an online community of college freshmen attempting to launch a government coup from the comforts of their dorm rooms. But they were mathematicians, computer scientists, privacy enthusiasts, lead scientists, future Turing Award winners—practically full-grown, men, and women who weaponized their technical expertise and political ideologies to inspire hundreds and then millions.

The movement was founded in 1992 by three men: UC Berkeley mathematician Eric Hughes, former chief scientist at Intel Timothy C. May, and Sun Microsystems computer scientist John Gilmore.[5] In the summer of 1992, after months of conversations about secure computing, cryptography, and kindred topics, the trio gathered 20 of their like-minded friends at Hughes' house in Oakland to discuss the future of technology, data privacy, the economy, and cryptography.[6] After a few months of meetings, the group adopted the name "Cypherpunks," a pun on the title Cyberpunk coined by famed hacker and founding member, Jude Milhon.

At the heart of the Cypherpunks' message were two essential questions: whether the government would use this exciting emerging technology called the internet to intrude on people's privacy with electronic surveillance, or whether the internet could break down international barriers and allow the world to communicate freely and anonymously.

With this in mind, the Cypherpunks championed individual privacy and aspired to create systems that allowed for barrierless interactions, anonymous transactions, and the employment of digital cash.

Where Did the Cypherpunks Come from Anyway?

There was still a missing piece of the puzzle—who or what inspired the Cypherpunks to form in the first place? Did Hughes et al. just wake up one day and decide to start a cyber gang, or was there something more? Interestingly, the origin of the Cypherpunks movement dates before 1992.

A similar crusade that began over a decade prior would set the course for the Cypherpunks and Bitcoin. It was a crusade set against the backdrop of the creation of public-key cryptography in the mid-seventies and the impending internet age in the late eighties. The individuals that led the crusade did not have a name like the Cypherpunks but consisted of a group of cryptologists and scientists that shared a common belief in cryptography's power to institute a society free from government coercion. At its core was legendary computer scientist and cryptography icon David Chaum.

David Chaum – The Original Cypherpunk

David L. Chaum was one of the most influential professionals in cryptology in the twentieth century. Chaum was not a founding member of the Cypherpunks, but his discoveries in privacy-enabling systems, electronic cash, and anonymous payment systems in the late seventies and eighties laid the technical groundwork for the Cypherpunks movement[7] and cryptocurrency.

The internet age of the seventies and eighties spurred a digital revolution. Back then, the internet was only available to government agencies and a handful of private companies, but that quickly changed. The internet was expanding to the private sector, and it was only a matter of time before private citizens used it to communicate and transact goods and services. Chaum believed that personal information was tied to everything—the people we talk to, the things we buy, and the places we go.

He postulated that, as the internet became a regular part of our day-to-day lives, the government and private institutions could use personal information collected from email and electronic payment systems to exploit people's lifestyles, habits, locations, and associations.[8] He warned that these unwanted infringements would eventually create an environment where people were afraid to express their political views, vote in the direction they wanted to vote, and safely talk to friends and family about matters of the heart. And because people would feel the need to self-censor, they would lose the ability to guide society in a direction that best served the people.

To prevent this inevitable outcome, Chaum spent his life discovering ways to use cryptography to enable society to evolve naturally.

Discoveries and Creations

Chaum was a young graduate student at UC Santa Barbara in the late seventies when he made his first significant contribution to the cryptography space. In 1979, he wrote his master's thesis, *Untraceable Email Systems, Return Addresses, and Digital Pseudonyms*[9]—an idea he conceived while sitting in a hot tub in a college professor's backyard—to discuss ways to solve society's privacy problems.[10]

He created *mix networks*, a cryptographic technique used to generate rosters of digital pseudonyms that the public could use to verify the accuracy of votes in an election.[11] His concept of mix networks later became a significant component in blockchain, Tim May's anonymous virtual environment, *Blacknet*, and the Cypherpunks' iconic mailing list. Email users could also use mix networks to hide their identities in emails without a third-party service to facilitate it. The scientific community praised his discovery, and it prompted them to look at cryptography's effect on privacy in a more profound way.

Chaum's endeavors in cryptology opened doors for greater discoveries. For example, in the early eighties when electronic banking services were growing in popularity, Chaum wrote "Blind Signatures for Untraceable Payments" (1983), a paper that first introduced the anonymous and untraceable payment system.[12] This novel payment system eventually became the heart of his digital money apparatus, *eCash*,[13] a computer-based currency that many consider the first cryptocurrency.

In the paper, Chaum held that as commercial electronic payment systems became more prominent in society, their use could reveal a lot of private information about customers to third-party payment providers, including things they buy, the places they go, the foods they eat, and even the charities they give to.[14] He proposed a new payment system that would permit customers to pay for items without disclosing personal information, what they bought, how they paid for it, when they paid for it, and the number of payments they made.

This may not sound very fascinating now that we live in an age where people can deposit checks, trade stocks, and pay for concert tickets with telephones they carry in their back pockets. However, if you look at it from the perspective of the period in which these concepts were proposed—the early-eighties when people still used VCRs, the World Wide Web didn't exist, and the Nintendo game console had not been invented—they were revolutionary.

The Encryption Dilemma—Shade from U.S. Agencies

Chaum's cryptography breakthrough came at a time when the U.S. government did not have a friendly outlook toward universities and corporations that studied cryptography.

Before 1996, the government held a monopoly on strong data encryption software. They primarily used it to prevent foreign adversaries from eavesdropping on their electronic communications. However, they discriminately obstructed American corporations and academic researchers from advancing encryption and using it for commercial purposes.[15]

In those days, strong data encryption software was classified as *arms, ammunition, and implements* under the International Traffic in Arms Regulations (ITAR) because of its capacity to facilitate anonymity.[16] The government was afraid that if researchers were to disclose cryptographic formulas to the public—whether through research publications, conferences, or conversation— that it would alert foreign adversaries of flaws in their coding system, thereby inducing them to change their codes, and causing the U.S. to lose all ability to eavesdrop on their communications. The government further feared that public cryptography could expose weaknesses in their own coding systems, leaving them vulnerable to exploitation by other nations.

The Department of State considered any disclosure of encryption information an "export of munitions" without a valid federal license and therefore a federal crime.[17] With that, the government threatened researchers with imprisonment if they intentionally broke this rule.

Chaum vs. the NSA

Chaum tells a story about his experience dealing with the NSA as a graduate student.[18] In the late seventies and eighties, the government issued several secrecy orders to cryptologists at universities and independent organizations, contending that disseminating research on cryptography either through written publication or verbally at conferences was a federal crime.[19]

For example, in 1977 the NSA wrote a letter to the Institute of Electrical and Electronics Engineers (IEEE) warning that their articles on cryptography and their upcoming cryptography conference at Cornell might violate ITAR.[20] Likewise, in 1978 researchers at MIT intended to publish a white paper on using public-key encryption to encrypt email messages, but because their discovery unleashed the potential for people and organizations to mask their communications from the federal eye, federal regulators used the same ITAR regulation to pressure the university not to publish it.[21]

Chaum, however, was confident that his discoveries and the discoveries made by his colleagues, especially those censored by the government, were the key to unlocking his vision for a free and utopian society. So, instead of muzzling his work, he responded by organizing a conference on cryptography, dubbed "Crypto '81", at his alma mater in August of 1981.[22]

He planned to invite as many friends and colleagues as possible to the conference but to do it in secret. Chaum was concerned that, because the NSA was threatening his colleagues with secrecy orders and potential imprisonment, the NSA was wiretapping the phones of known cryptologists, including his own. So, instead of making invites over the phone, he went old school and invited them in person or through the U.S mail. About 100 attendees[23] from all over the country, as well as scientists from Canada, England, and Sweden,[24] turned out for the three-day conference. In Chaum's own words, "Most people interested in the field came out to it."[25]

Crypto '81's success inspired Chaum and his colleagues to form the International Association for Cryptologic Research (IACR), a non-profit organization devoted to promoting cryptology that is still thriving today.[26]

Chaum went on to make other significant breakthroughs in cryptography and digital currency, including the creation of Digicash, the world's first digital cash payment system,[27] which is touted as a precursor to Bitcoin.

The Voice of the Union

Chaum's breakthroughs in cryptography and fervent push for increased privacy in society inspired an entire culture of cryptologists and philosophers in the eighties, including Tim May. You met May a few sections ago; he was one of the co-founders of the Cypherpunks.

In his 1988 paper, "The Cryptoanarchist Manifesto," May argued that, with technologies like public-key encryption, zero-knowledge proof systems, and applicative software protocols, interactions between individuals and groups would be impervious to government tampering and surveillance.[28] He believed these systems enabled people to choose the amount (and type) of information they wanted to disclose to others. Obviously with tremendous foresight, May predicted that people would someday use encryption to protect their identity—in email, messaging apps, computer and phone PIN/password, financial transactions, etcetera—in an era where the internet was not yet available to the public. Four years later, the Cypherpunks were born.

Cypherpunks Unite – the Birth of the Cypherpunks Mailing List

The Cypherpunks knew there were people out there who believed in the cause, but they needed a way to reach them. They created a mailing list that people who believed in the cause could subscribe to and share ideas on privacy, anonymity, digital cash, cryptology, philosophy, politics, government monitoring, and corporate control of data.[29] The mailing list propagated various websites for scientists, cryptologists, coders, and other like-minded individuals, and within the first few weeks, it received one hundred subscribers.

The mailing list was one step in promulgating the Cypherpunks' message, but there was still the problem of anonymity. They wanted subscribers to share ideas freely without fearing reprisal from the authorities. To prevent this Eric Hughes and Hal Finney created an anonymous remailer using Chaum's concept of *mix networks*.[30] The remailer allowed a sender to send emails to a recipient or a chain of recipients in such a way that the email could not be traced back to the original sender. In some respects, it was an early representation of a blockchain as the remailer operated on a distributed network using public-key cryptography.

Unstoppables

The Cypherpunks were on the move. They'd evolved from a small social group to a worldwide movement with branches established in Europe,[31] and Hughes was smart about it. He observed how positively their fans in cyberspace responded to their message, so he made it official.

In 1993, Hughes published the Cypherpunks' official mission statement, "A Cypherpunk's Manifesto." In it, he made a case for society's need for a system where people could freely choose the amount of personal information they wanted disclosed to others.[32] He and the other Cypherpunks believed that the best way to achieve this was to combine distinct components of technologies: cryptography, anonymous mail-forwarding systems, digital signatures, and electronic money. The Manifesto reached hundreds. Their message got louder, and their mailing list grew. By 1994, the mailing list reached 700 subscribers,[33] and by 1997 it had reached 2,000.[34]

Home Stretch – the Road to Bitcoin

It's fascinating to see how the early Cypherpunks borrowed various technical components from David Chaum's discoveries to build their unique versions of an anonymous electronic cash payment system. It's even more fascinating, as you will soon see, how Satoshi Nakamoto borrowed technical components from various Cypherpunk electronic cash payment systems to create Bitcoin. From this perspective you could almost say that Bitcoin is the child of the Cypherpunks and the grandchild of David Chaum.

But the Cypherpunks of the late nineties and early 2000s kept Chaum's dream of digital cash alive. Like Bitcoin, many of the electronic cash payment system proposals and innovations that came after Chaum's eCash invention were built with applications designed to mitigate inefficiencies in the financial system—double spending, fraud, theft, corruption, and the need to trust a third party.

- **Hashcash (1997) – Proof-of-Work and Timestamps.** Circling back to the explosive growth of the Cypherpunks' mailing list, the Cypherpunks realized that not even their super-anonymous email system was impervious to abuse. As their mailing list grew, so did the number of spammers that flooded the system. Spam bombs became a big problem in the mailing list as some users would receive hundreds of spam messages in a single day.[35] The Cypherpunks wanted their members to focus on legit emails without all the distractions from spam, so they sought a solution. That's when it happened.

  In 1997, now-Blockstream CEO and Cypherpunk OG, Adam Back, developed a brilliant solution that required computers (computers owned by the users sending the emails) to solve a cryptographic puzzle before they were allowed to send an email. He named this particular action, *Hashcash*.[36]

  The puzzles were not easy. Computers had to do difficult and time-consuming computations to discover the correct answer.

Computers essentially had to *prove* that they did the *work* required to solve the puzzle. As a result, users received a *hash token*—a single string of letters and numbers that proved that the work was complete.

Example hash token:
00000002YH4KSB03PNQ440N4G-
DLF035F1D7OM9WQA4

That token was their pass to send an email. This entire process required not only time but computing power. And computing power costs money. Hashcash was so effective in blocking out spam because spammers could not afford the computing power necessary to send out hundreds of emails, so they gave up.

Hashcash was one of the first known applications of the proof-of-work process, which a decade later would become the primary computer process used to mine bitcoin. We will discuss the awesomeness of Bitcoin mining in future chapters.

It's worth noting that Adam Back's Hashcash was so valuable in Bitcoin's creations that it was cited in Bitcoin's 2008 white paper.[37] (With respect to time stamps, I should also mention that Bellcore Labs' research scientists Stuart Haber and Scott Stornetta's 1991 paper, "How to Time-Stamp a Digital Document," was also a crucial part of Bitcoin's time-stamping application and blockchain in general.)

■ **b-Money (1998) – Decentralized Ledgers, Broadcasting, and Mining Incentives.** In 1998, a computer engineer named Wei Dai created a proposal for a digital currency called b-Money.[38] The b-Money runs on a network that enables users to trade b-Money with each other in an anonymous way. Dai created applications for b-Money that Satoshi Nakamoto later adopted to create Bitcoin, including decentralized ledgers, broadcasting messages and signatures, and incentivizing miners to mine bitcoin.[39] Dai, too, was cited in Bitcoin's white paper.

- **Bit Gold (1998 – (Full proposal released in 2005)) – Proof-of-Work and Registries.** Nick Szabo, a former employee of Chaum's and the creator of our beloved smart contracts, created a proposal for *Bit Gold*, a decentralized digital currency that utilized the proof-of-work consensus processes, digital timestamps, and a publicly verifiable registry.[40] Satoshi Nakamoto confirmed Bit Gold's influence on Bitcoin in a 2010 Bitcoin forum post:

Bitcoin is an implementation of Wei Dai's b-money proposal http:// weidai.com/bmoney.txt on Cypherpunks http://en.wikipedia.org/wiki/ Cypherpunks in 1998 and Nick Szabo's Bitgold proposal http://unenumerated.blogspot.com/2005/12/bit-gold.html [41]

—Satoshi Nakamoto

- **RPOW (2004) – Reusable Tokens.** In 2004, legendary cryptographer and Cypherpunk member Hal Finney created the Reusable Proofs of Work or *RPOW*.[42] RPOW was an improvement on Adam Back's Hashcash and then some. With Hashcash, computers generated tokens that could only be used once (to send an email). With an RPOW token, however, once it's used (transacted from person-to-person), it will automatically generate a new one in its place, which can be used again (and again and again).[43] Not only that, but RPOW was the first public implementation of a server designed to allow users throughout the world to verify its correctness and integrity in real-time. I think it's worth noting that Finney received the first Bitcoin transaction in 2009.[44]

The Cypherpunks movement dwindled in the early 2000s. After 9/11, the U.S. government heightened surveillance on electronic communications, including online chat forums. Many of the original Cypherpunks' mailing list subscribers feared being targeted by the government so they unsubscribed.[45] Likewise, many of the Cypherpunks' founding members pursued separate ventures in the cryptography space (including some of the proposals discussed in the previous section).

But the movement for digital privacy was far from over. In 2007, a new mailing list called the *Cryptography Mailing List* appeared and continued the conversation that had fizzled out with the end of the Cypherpunks' mailing list. The focus of this new mailing list, however, was primarily on the prospects of digital currency.[46] Little did people know that on one fateful day in the fall of 2008, a subscriber going by the screen name *Satoshi Nakamoto* would post a nine-page white paper on a digital currency so revolutionary that it would upend the traditional financial system and bring financial hope to millions.

CHAPTER TWO

BITCOIN
Striving for Greatness
in the Face of Adversity

The financial crisis of 2007–2009 brought America (and nearly the rest of the world) to its knees. Most of you lived through it and can remember its effects on the economy and maybe your own lives: neighborhoods full of homes with "Foreclosed" signs on wooden steaks in their front yards; files of laid-off workers marching out of office buildings with personal belongings in hand; interest rates on credit cards shooting up to highs that were barely below the federal legal limit. I vividly remember Federal Reserve chairman (1987–2006), Alan Greenspan, declaring before Congress, *"I made a mistake in presuming that the self-interest of organizations, specifically banks, is such that they were best capable of protecting shareholders and equity in the firms."*[47] Nevertheless, it was too late. The country had already fallen into a pit of despair and had nowhere to look but up.

Recap of The Financial Crisis of 2008

For those of you who were not around during the financial crisis, allow me to briefly recap. Decreasing interest rates for "prime" bank customers—that is, customers that banks considered low risk—propelled major Wall Street banks to reinvigorate a 20-something-year-old money-making scheme whereby banks offered low-interest adjustable-rate mortgages to customers that were usually deemed as "high risk" or "subprime." These loans were labeled subprime mortgages.

Subprime mortgages were just one of many mortgage-backed securities products banks offered to investors who wanted to profit from the booming housing market. Banks approved subprime mortgage loans by the millions. As a result, the demand for housing increased. Hence, banks continued to supply high-risk customers with enough capital to purchase one, two, or even three consecutive properties, sometimes without conducting credit checks or verifying their employment.

Housing prices spiked to record highs, and banks and investors alike amassed billions of dollars from mortgage sales and interest payments. What banks (or anyone else) did not anticipate, however, was that interest rates would once again rise. Home buyers no longer received good deals on interest rates, so the demand for homes gradually declined along with the homes' value. And because banks initially supplied most high-risk customers with adjustable-rate mortgage loans, loan payments soared to prices far above what homeowners could afford, so they defaulted. The culmination of mortgage defaults led to a nationwide home foreclosure pandemic. What followed was the nation's worst recession since the Great Depression.

Financial institutions and federal regulators (and we Americans) were caught off guard by the rapid deterioration of the financial system, but signs of a recession had started years before the recession fully materialized.

In 2006, housing prices fell for the first time in eleven years as borrowers defaulted on their loans.[48] Some banks didn't reserve enough capital to support the crushing weight of the defaults, so by 2007 banks began to fail. In March and April of 2007, more than 25 subprime lending firms—including one of the nation's premier mortgage banks, New Century Financial Corporation—declared bankruptcy.[49]

The sudden collapse of mortgage lenders spooked investors. Many looked on as the subprime mortgage sector capsized and scrabbled to rescue their funds from the vacuum before they were swallowed up.

The Federal Reserve took action to slow the hemorrhaging by making a series of rate cuts, but the decline continued to accelerate. More homeowners defaulted on their loans, more businesses shut their doors, and Wall Street institutions that were once thought of as invincible fell within weeks.

In March 2008, Bear Stearns, one of the largest and most dependable investment banks on Wall Street, tanked. It was subsequently acquired by J. P. Morgan Chase for only $10 per share (the stock was worth over $100 at the previous close). By September, Lehman

Brothers, one of the largest global investment banks, collapsed in like manner. Mortgage financers Freddie Mac and Fannie Mae, which owned or guaranteed nearly half the mortgages in all of the U.S.,[50] were on the brink of collapsing, but the U.S. government stepped in with a bailout.

Because these two institutions were so nested into the fabric of the global financial system, the government feared that if they allowed them to fail, economies worldwide would be decimated. Likewise, giant bailouts were granted to banks like Wells Fargo and to finance and insurance corporations like American International Group (AIG) that were "deemed too big to fail."[51]

Looking back at the fallout, the final numbers are sobering: in a little over a year, the housing market dropped 30%,[52] the stock market lost nearly $8 trillion in value, unemployment hit a peak of 10% (October 2009), retirement accounts lost $2.8 trillion in value, and the government spent upwards of $440 billion on bailouts.[53]

I said all of that because I wanted to paint a vivid picture of the demise of these institutions. Why? Because the way these institutions went under was once thought of as impossible. No one ever imagined that mega financial institutions that had existed on Wall Street for 80-plus years could fall in a few months. Banks and investors were confident that the housing market was a safe investment, immune from collapse. Banks were not deterred from approving risky loans because they planned on selling them to the mortgage companies on Wall Street anyway.

Likewise, mortgage companies were not concerned with purchasing risky loans because the housing market was thriving and they were promised that their credit default swap contracts would protect them from losses. They were both wrong. The banks concocted a flawed money-making system that resulted in millions of Americans becoming jobless and/or homeless.

The financial crisis proved to the world that financial institutions and the financial systems are heavily defective and that there needed to exist a financial system separate from the state and not based on trust. And like clockwork, along came Bitcoin.

Digital Cash World Order

Bitcoin was created against the backdrop of the 2008 financial crisis. Americans felt the crushing weight of debt, foreclosure, joblessness, and the hopelessness of society. They felt the betrayal of the nation's most trusted financial institutions. And though Bitcoin was not created in response to the financial crisis, it showed the world that a financial system separate from the existing flawed financial system was possible.

Bitcoin was initially designed as a payment system, but the financial crisis quickly prompted early admirers of the project to underscore its other pragmatic uses, such as its ability to enable cheap cross-border transactions and its capacity to be used as an investment instrument to gain wealth.

At the time, these were qualities that the nation—and frankly the world—coveted because they eliminated society's reliance on the state's failed financial system and put the monetary control into the hands of consumers.

But if the state of the economy necessitated Bitcoin, then why didn't people warm up to it? We will look at possible reasons for people's aversion to Bitcoin.

Though Bitcoin appeared to be a reputable alternative to the old system, its adoption did not happen immediately. People were reluctant to trust Bitcoin for several reasons, three of them were: the inability to fathom using a new type of currency, the obscurity of how Bitcoin obtained its value, and the excess misinformation about Bitcoin.

Bitcoin, the Currency Evolution

Bitcoin was a foreign concept, a new form of currency that could not be physically held or spent but was created by a computer code, something that most people did not think was possible. It was undoubtedly a complex concept for people to comprehend, given that the world has been using paper currency for the last few centuries.

Let's think about what currency is for a moment. People have used some form of currency for over ten thousand years. Before people used paper currency and metal coins, they used other items that society at the time considered valuable to trade for goods and services. Cattle are the oldest known form of money.

Since at least 9000 BC, people bartered for goods and services with cows, sheep, goats, oxen, and other livestock. In biblical times, a person's wealth was measured by the number of cattle he owned.[54] In 1300 BC, the Chinese used cowrie shells (a small, egg-shaped marine mollusk found in the Indian Ocean) as currency.[55] Imitation versions of the cowrie shell made from copper and bronze, the earliest form of metal money, were adopted around 1000 BC in China and then expanded into Africa. With time, those egg-shaped metal cowrie imitations took a round shape, like a coin. And in AD 800, paper currency was born in China (paper currency was eliminated there around AD 1450 but reemerged in Europe in the seventeenth century).[56]

As you can see, new currencies introduced to the world are nothing new. In fact, the most significant introduction of a new currency in history[57] occurred as recently as 2002. On January 1, 2002, 12 countries in Europe ditched their currencies (some of them centuries old) in lieu of the euro. Today, over 20 years later, around 341 million people use it every day, making it the second most used currency worldwide, according to European Union data.[58]

In light of these facts, introducing Bitcoin as a new type of currency doesn't seem as farfetched, right? People will probably be more open to Bitcoin if they understand that money is an evolutionary element, and the shape, style, and material of money will vary widely.

Bitcoin Does Have Value

Crypto critics have a hard time identifying the value of Bitcoin, and they are not afraid to say so. They say it has no value because they can't hold it in their hands or the government doesn't recognize it as official currency. This is understandable, given that society has mainly interacted with assets in the physical realm for millennia. But people can sometimes overthink how value is actually determined. They look for a mathematical explanation for how an item derives its value when, in fact, it's determined simply by what people say the value is.

Think of an antique painting. Why do people pay hundreds of thousands—and sometimes millions—of dollars for a single painting? At its core, it's only made of canvas (or wood) and paint, right? But the market discovers its value based on the painting's originality, the period it was created, its scarcity, what it represents, and the status it gives the owner. Some collectors see a work of art as an appreciating asset like a house or a stock.

You can't follow arguments about Bitcoin's value for long without hearing the term *intrinsic value*. The notion of intrinsic value has spurred a lot of philosophical debate about how to define and properly apply the term, especially to Bitcoin (and other cryptocurrencies).

I want to demystify some things about intrinsic value that I think people get confused about when applying the term.

There are two types of intrinsic value people apply depending on the subject. One type is used to calculate the value of a stock or company, which involves complex calculations and considers variables like trademarks, copyrights, the quality of the directors, brand name,[59] etcetera. The other is the base definition of the term, something you will find in a legal dictionary. Since neither fiat currency nor Bitcoin is a stock or company, we can logically assume that when people talk about the *intrinsic value* of Bitcoin, they are referring to its book definition:

Something is intrinsically valuable when its value is tied up within itself. In other words, it's valuable for its own sake.

For example, virtues like joy, happiness, health, and love have intrinsic value because they are valuable states of being within themselves. People want to *be* joyful; they want to *be* happy; they want to *be* healthy; they want to *be* loved. They do not want these things just to have them; they want to *be* them.

If you apply this meaning to assets like fiat currency, gold, and Bitcoin, you will see that they have no *intrinsic* value because they are within themselves useless. But that is not to say that they do not have value. They do.

Let's look at a different kind of value that I believe accurately applies: *instrumental value*. Something has instrumental value when people use it as a means to an end. For example, people use money, gold, Bitcoin and other assets to get something they value, whether it be more possessions or financial safety. But in contrast with things valued intrinsically, people want instrumentally valuable things for the sake of having them.

Now, given the facts of instrumental value, consider what people use Bitcoin for. They use it to pay for goods and services, make (partially) anonymous transactions, send fast and cheap borderless transactions to family and friends around the world, as an investment vehicle to grow wealth, and to obtain financial independence from the oppressive and discriminatory financial systems. This shows that Bitcoin is a means to many different ends and therefore has value.

Bitcoin #FakeNews

Misinformation is false information spread by a person or group of people that is widely believed but largely incorrect.

There was a lot of misinformation circulating about Bitcoin when it started to go mainstream in 2011, and that turned people against it.

I remember reading a blog comment in 2013 where a guy wrote, *"You guys are amateurs and you have no idea what the [expletive] you're doing. Don't daytrade this [Bitcoin], don't use a "buy and hold" strategy. Don't do anything because you dont KNOW anything. We literally spend a good part of the day making fun of you guys…Tulip mania, look it up, and then cash out and be thankful you have anything left in your account"*

Raise your hand if you've heard any of the following?

"Bitcoin is a Ponzi Scheme," "Bitcoin is a bubble," "Bitcoin is like gambling," "No one will ever accept bitcoin as payment," "Bitcoin is only speculation," "Bitcoin has no real value," "Bitcoin is too hard people to understand," "Only criminals will only use Bitcoin"

We've heard it all.

Pinpointing the root of the misinformation is moot at this point, but its effect influenced people's decisions to learn about or invest in Bitcoin. It was infinitely easier to believe misinformation about Bitcoin because it was a novel concept that was contrary to the way people experienced technology and finance. This made the misinformation more dangerous.

As a personal example, every now and then a former coworker and I would get together to talk about the stock market and occasionally share a stock tip or two. I told him about bitcoin, and that he should consider buying some because the trading momentum was picking up and it had started to get a little mainstream attention. Now, my coworker *knew about* cryptocurrency, but he didn't *know* cryptocurrency. He knew very little about how it functioned and how to trade it—he knew only as much as he'd heard about it from the skeptics in his life.

In the beginning, he'd shrug it off and switch the topic back to stocks, but I continued to bring it up after it hit a new milestone. I pitched Bitcoin to him for a solid year, and he would comment with something like, "Cryptocurrency is not going anywhere" or "Cryptocurrency will be too hard for regular people to understand; it doesn't have a chance for survival." I kept it up for a while longer but decided to give up after he said that bitcoin would be so regulated that no one will make money from it.

Needless to say, he never took my advice, and we now know what Bitcoin became, but it goes to show you that misinformation can have severe consequences on a person's perception.

The Psychology of Hating Bitcoin

There is a psychological explanation for why people are hesitant to adopt Bitcoin despite the country's turmoil during the financial crisis. People become skeptical about accepting new technology into their lives, but the motivations behind their reluctance often go unnoticed. Is there a why behind the why? When you peel back the layers, most of it is rooted in fear and perception.

The Fear

Fear is an innate human emotion. It alerts us to the presence of perceived danger, whether real or imagined.

You've probably heard of the famous idiom *Fight, Flight, or Freeze.* We tend to respond to fear in one of these ways. We avoid it *(Flight)*— which is our preferred response if a situation permits us to walk away or avoid it altogether; we lash out physically or verbally *(Fight)*—which is how we react if backed into a corner; or we become disoriented *(Freeze)*—which is not the best response to fear because it leaves us vulnerable to that feared situation. With Bitcoin, we saw that people respond by avoiding it entirely or lashing out against it on social media, in news articles, and live interviews.

It's important to note, however, that people's hesitancy towards Bitcoin had nothing to do with Bitcoin per se. People were not afraid of the actual coin—like one would be frightened by a snake or a spider—but were fearful of feeling unpleasant emotions if their experience with Bitcoin was negative. Here is a list of common fears that prevent people from being involved with Bitcoin:

People are…

- Afraid that the process of buying bitcoin exceeded their technical proficiency, thereby making them look or feel inadequate.

- Afraid to step out into something new and feel safe living in the status quo.

- Afraid that the probability of losing money was too high.

- Afraid that they could not bear the emotional unpleasantness of falling for a scam or being taken advantage of.

- Afraid that the responsibilities associated with investing (stressing over the market, managing multiple investments, putting more effort into learning about the space, fearing being hacked, etcetera) were too stressful for them to handle.

Fear can indeed be unsettling, but not all fear is bad. Fear serves a vital role in keeping us safe from harm. But if left unchecked, fear can make our world small and keep our options limited.

First Impressions

Have you ever gone to a restaurant or coffee shop to meet a lady for a first date, but she arrived fifteen minutes late without giving any kind of explanation or apology? And when she finally arrived she was drunk? Or maybe she didn't look quite like the person in her profile photo? If so, you probably didn't feel too hot about that and had already decided never to see her again.

People say that first impressions are everything. Social psychologists have studied this phenomenon for decades and found that we (people) unconsciously consider a person's physical attractiveness,[60] verbal and nonverbal expression,[61] behaviors, race, gender, speech patterns, location of residence, etcetera when forming our attitudes toward that person, particularly to assess their trustworthiness and competence. Some first impressions can last months and can cause people to become biased even when there is future evidence that contradicts initial judgments.[62]

I think that these concepts can also apply to non-human objects. For example, have you ever decided not to buy an item from Amazon because you did not like some of its customer reviews? Or maybe there was a particular movie you wanted to see but decided not to because it only scored 19% on Rotten Tomatoes? Whether we realize it or not, the opinion of others can shape our own, especially if we have no familiarity with the item or object in question.

Some people became an enemy of Bitcoin when their first exposure to it was a negative report, news article, or personal testimony from a friend or family member. And they may have imparted those negative impressions to others, and those people to others, and so on. Before long, misinformation about Bitcoin has touched tens, hundreds, or even thousands of people.

As with fear, people can be blind to how others influence their opinions. For example, people who called Bitcoin a scam or valueless may have never read the Bitcoin white paper or tried to understand how the technology worked or looked at how it served the unbanked in impoverished communities or read an article on how it financially liberated people living under oppressive regimes in Africa and South America, or compared its advantages and disadvantages to the advantages and disadvantages of the existing financial system. They simply perpetuated what they *heard* someone say or *read* on social media.

Unfortunately, once our perceptions are tarnished, it's not easy to consider the other side of the argument. That's because we tend to focus on information that confirms our pre-existing judgments and dismiss information that contradicts those judgments.[63] This may explain why people who hated Bitcoin back in the day still hate Bitcoin despite its mainstream adoption and success.

The Tide Has Shifted

Until 2020, banks all but avoided crypto. But as you will see in the following chapters, some banks were so afraid of cryptocurrency that they froze or closed the bank accounts of crypto traders whose accounts were linked to crypto exchanges. But, since the crypto boom of 2020–2021, multiple banks and financial services have offered cryptocurrency investment services to their customers.

For example, in America, several financial companies filed spot Bit-coin Exchange Traded Funds (ETF) applications with the U.S. Securities and Exchange Commission (SEC), while companies like Grayscale, Valkyrie, Vaneck, and more began giving customers exposure to Bitcoin through futures-based cryptocurrency ETFs. Valkyrie and Osprey even offer clients trusts for Polkadot, a project not as widely known as Bitcoin but desired by mainstream investors.

In April 2022, Fidelity announced that it would offer investors a way to put bitcoin into their 401(k)s, the first retirement plan provider to do so.[64] Some financial institutions like Morgan Stanley,[65] J. P. Morgan,[66] and Bank of America[67] stood up cryptocurrency and blockchain research divisions to explore opportunities in the space.

The tide has shifted. People have warmed up to Bitcoin. But why now? After all, the world wasn't suffering from a financial crisis then, right? In fact, they were. The COVID-19 pandemic of 2020 caused the worst financial crisis since 2008.[68] The Dow Jones Industrial Average fell 26%,[69] thousands of small businesses shut their doors, millions of people were laid off, and the government spent trillions of dollars on economic recovery plans. However, this time there was a combination of factors that made Bitcoin stand out as a reliable alternative:

-Cryptocurrency had 10 years to gain traction.

-People were making a lot of money—which reeled in new retail in-vestors and mainstream financial institutions.

-Billions of people worldwide were quarantined so they had time to seek other endeavors like developing crypto projects, reading news articles, or watching YouTube videos about cryptocurrency, etcetera.

Our Refuge

Bitcoin is now a part of life. You can send, receive, and spend it through third-party payment providers; stores, shops, restaurants, and gas stations around the world accept it; and mainstream financial institutions offer it to clients. Some countries have even adopted it as legal tender. Bitcoin has achieved more than most people would have imagined in such a short time. But there is still a long way to go. The crypto-community is making strides in promoting Bitcoin not only as an investment instrument and a medium of exchange, but also as a haven for those who the traditional financial system has failed/is failing. Time will tell if another financial crisis like 2008 will hit. If it does, I hope people will know they can always take refuge in Bitcoin.

CHAPTER THREE

THE ELON EFFECT
Could One Person Cripple the Crypto Market?

Question: Who has $260 billion, owns a suite of sweet cars and rockets, and can single-handedly break the crypto market with just one tweet? Yep, you guessed it, your favorite billionaire, Elon Musk.

Does Elon even need an introduction? Born in South Africa, co-founded PayPal, owns Tesla and SpaceX, etcetera, etcetera. The conventional world might know him as the guy who makes cool cars and sends rockets into outer space, but people in the crypto world know him as the guy who can move billions of dollars in or out of the crypto market with just one tweet.

The crypto market in 2021 was an emotional roller coaster ride for crypto investors, and for a while, Elon was operating the controls. When my phone pinged with news alerts about bitcoin or Dogecoin reaching a new high or falling to an all-new low, part of me assumed Elon's tweeting had something to do with it even before I knew what triggered the price move. Some described his tweeting as *reckless*— posting the first thing that came into his head about cryptocurrency with no regard to how it would affect investors; others thought he was purposely using his influence to manipulate the crypto market for his own gain.

But despite the rumors and accusations, Elon's involvement in Bitcoin was key to the crypto boom, so much so that the crypto-community made Elon an unofficial figurehead of Bitcoin and Dogecoin.

I need to make one thing clear before we move on. I gave Elon his chapter in this book, but I am in no way suggesting that he throws shade on cryptocurrency. On the contrary, he is good for cryptocurrency in many ways because he can bridge the crypto world to the non-crypto world, leading to further (and faster) cryptocurrency adoption. But I believe that his influence on the crypto market sets a dangerous precedence that, if wielded by the wrong person, can result in malicious misinformation, manipulation, and destruction. That is something the crypto market cannot allow.

Life Before the Crypto Boom

Before the crypto boom there was barely any mainstream news coverage on cryptocurrency. We were dead smack in the middle of a nearly two-year crypto-winter and traders were passively watching the crypto market for signs of life. CNBC might mention bitcoin a few times, the price would spike, and three days later, bitcoin corrected and returned to its previous high. Nothing exciting. However, momentum grew in mid-2019, and some financial news agencies took notice.

Yahoo! Finance had an entire page devoted to cryptocurrency price movements. It only had a few hundred coins listed (compared to the 8,000 there are today), but to those of us who'd been in the crypto game for a while, it was a sign that cryptocurrency was gaining momentum in areas outside of the crypto-community. Eventually, other financial news agencies caught on. Some even posted one or two news stories about cryptocurrency on the same page as the equity markets and bond markets. Whether they knew it or not, they were putting the crypto market on equal footing with the broader financial markets, and to us in the crypto-community, it was a good thing.

Something big was happening. It was something that could not be seen but was felt. It took time, but the world eventually woke up to cryptocurrency, and in 2020, things got interesting.

Independence Day

Barring chatter about bitcoin and altcoins on crypto Twitter, the crypto market stayed relatively quiet for the first three quarters of 2020. Bitcoin's price fluctuated between $6,000 and $10,000 per coin. If you were trading crypto at the time, you'll remember some of the social media debates on whether bitcoin had any chance to rise higher.

But stuff got real in Q4. We were six months into the COVID-19 pandemic and daily updates of COVID-related deaths rang from every news outlet worldwide. COVID-19's effect on U.S., Asia, and Europe stock markets resulted in individual investors, stock-traded companies, and even retirees losing capital.[70] In the U.S., the economic fallout caused the Federal Reserve to cut rates and print money as fast as they could count it. Retail and institutional investors sought refuge in alternative finance—something impervious to inflation and a weakening dollar—and that alternative was Bitcoin.

It was finally happening. For the first time, mainstream institutional investors were stockpiling bitcoin. MicroStrategy, Square, Mass-Mutual, Skybridge Capital—all these companies collectively jumped into the Bitcoin market, causing it to explode like fireworks on the Fourth of July. Bitcoin's price skyrocketed from $10,795 at the beginning of October to a high of $29,244 in late December.[71] It was one of the most significant bitcoin price surges in its history.

Words cannot explain how validating it was to the crypto-community when multi-million and billion-dollar companies invested in Bitcoin. Some of you know what I'm talking about. Crypto advocates spent years advocating for mainstream cryptocurrency adoption only to get shut down and laughed at. But now it was happening right before their eyes.

It was Bitcoin's time to prove that it deserved a position in the financial system.

Enter Elon Musk, the Twitter King

The momentum from 2020 carried over into 2021. Bitcoin's rise slowed a bit, but no one was complaining; it was up more than anyone could anticipate, and the world was finally paying attention. Likewise, altcoins like Ether, VET, Maker, Litecoin, and many more were up too, hundreds of percent. Crypto investors thought they'd seen bitcoin perform at its best, that is until Elon started tweeting.

In late January, Elon changed the bio portion of his Twitter account to *#bitcoin* (which creates the orange bitcoin symbol). That one change caused one of the highest price movements in 2021. Within hours, bitcoin's price jumped from $32,000 to $38,000, adding roughly $111 billion to Bitcoin's market cap.[72] That is a 17% price jump, which is rare if you've been trading cryptocurrency for a while.

In early February 2021, Tesla announced that it bought $1.5 billion worth of bitcoin and would start accepting Bitcoin to pay for some Tesla items.[73] Elon's move not only pushed the price of bitcoin up 16% that same day, but it set off a chain reaction of investors storming the crypto market. This officially began crypto 2021.

Tesla's announcement was a pivotal moment for cryptocurrency. Elon was essentially telling the world that bitcoin was solid enough to compete with fiat currencies.

Retail investors started buying up coins like it was going out of style and institutional investors like Morgan Stanley and BNY Mellon poured into the market. Cryptocurrency exchange Coinbase went public toward the end of the year, and bitcoin blew past an all-time high of $60,000 per coin. Every single cryptocurrency made daily gains, some in the hundreds of percent. The crypto market was so lucrative that people who never considered investing (in anything) sought ways to get started.

As far as the crypto-community was concerned, Elon was more than welcome to use his platform to bring mainstream awareness to Bitcoin and cryptocurrency in general.

But, We Celebrated Too Soon

Who knew Twitter (more like the man sending the tweets) could be so dangerous? In April and May of 2021, Tesla dropped two bombs on the crypto-community that brought the crypto-market euphoria to a violent halt. In April, Tesla's CFO Zach Kirkhorn announced during an after-earning call that Tesla sold 10% of its bitcoin (netting approximately $272 million[74]).

And if crypto traders weren't holding onto their dwindling bitcoin portfolios enough, the following month, Elon posed a statement from Tesla on Twitter announcing that Tesla would stop accepting bitcoin as payment for Tesla vehicles due to environmental concerns.[75]

The news sent the crypto market into a freefall. Bitcoin's price plunged by more than 12% after the tweet[76] and over 28% the following week after more ambiguous Bitcoin tweets from Elon. Billions of dollars were slashed from Bitcoin's market cap, and millions slashed from investors' portfolios.

It was evident by the number of F-bombs and sarcastic responses to Elon's post that the crypto-community's sentiment toward him (and Tesla) shifted from honor and adoration to betrayal and suspicion. I don't think anyone saw it coming. Even I was left with the proverbial WTF facial expression when I read about it. But to Elon's credit, he clarified that he did not sell any of his own Bitcoin, but Tesla sold 10% of its bitcoin to prove liquidity, which is common in business.[77] Yet the crypto-market wasn't so forgiving, and the volatility continued.

At that moment, the crypto-community may have relied too much on Elon's ability to thrust bitcoin (and other cryptocurrencies) to the moon. On some subconscious level, I think everyone believed that because Elon took a personal interest in Bitcoin (and tweeted about it all the time), he would not rush to make a decision that would affect the entire market. Everyone's gotten so accustomed to his showmanship that they've forgotten that he was first a businessman and that his entire business model is built around providing clean energy to the world.

But if anything, Tesla's stance on the environment ignited a heated worldwide debate about Bitcoin mining's potential to affect the environment negatively, so much so that in June of 2021, U.S. crypto mining companies, MicroStrategy CEO Michael Saylor, and Elon created the Bitcoin Mining Council to promote transparency, share best practices, and educate the public on the benefits of Bitcoin and Bitcoin mining.[78] The mining/environment conversation has moved nations to crack down on Bitcoin mining or ban Bitcoin altogether. We will look more closely at worldwide Bitcoin mining bans in a later chapter.

An Emoji Is Worth a Thousand Words

They say that pictures are worth a thousand words, but what about emojis? It turns out that the crypto-community believes they are.

The Tesla thing was still fresh on everyone's mind, but no one gave up on Bitcoin. We were still in a bull market and believed that Bitcoin would make a strong comeback, with or without Elon's help.

Just as the crypto market started to rebound, Elon took to Twitter and did something strange. On an early June morning, he tweeted *#bitcoin* with a broken heart emoji and a meme of a guy and girl looking like they were about to break up.[79]

People had their own theories on what the cryptic tweet meant:

1. *It signified bitcoin's violent drop before the rebound broke Elon's heart (he caused it).*

2. *Elon gets the sad when he's not tweeting about bitcoin.*

3. *Or, as far as 100% of the crypto market was concerned, Elon's love affair with Bitcoin was over and he was moving on to new things.*

Whatever the tweets' true meaning was, Bitcoin's price fell 7%[80] that morning. Later Elon tweeted a string of photos presumably poking fun at the massive market drop.[81] No one knew where he was going with any of his tweets, but the uncertainty crippled crypto investors with anxiety.

Is There Such a Thing as Having too Much Influence Over the Crypto Market?

The crypto-community had mixed feelings about how the market reacted after Elon's tweet. Some believed bitcoin's price plunge signified an immature market—a market of amateur investors that based their trading behaviors on emotions rather than sound expertise—and that it left the market vulnerable to manipulation. Others said that investors made the right move in pulling their funds out of bitcoin before they lost it all.

The crypto market's reaction also raised another, more important question: Should one person have the power to influence the market in a major way? I think the answer is obvious. Elon will only have as much influence as people give him.

Another vital question to consider is why Elon has a persuasive hold over some people. I believe it's because his brilliance, innovations, status, and vision inspire a new generation of entrepreneurs, tech enthusiasts, and clean energy advocates that want to usher in a new world with him. Let's not forget that he is also a cool, down-to-earth dude who connects to people personally. It's only natural that people will follow his lead.

The Dogefather

> *"Dogecoin might be my fav cryptocurrency. It's pretty cool"*
>   —Elon Musk, Twitter, April 2, 2019[82]

That was the tweet that started it all. What came after it was two years of tweets, debates, interviews, and guest appearances from the self-proclaimed *Dogefather*[83] promoting Dogecoin as the payment network that would one day take over the traditional financial system.[84]

Firstly, if you are not already familiar with Dogecoin, it is a cryptocurrency payment system created in 2013 by cryptocurrency developers Jackson Palmer and Billy Markus.[85] Dogecoin and its native cryptocurrency *DOGE* were initially created as a joke, poking fun at a popular internet meme featuring the Shiba Inu (Japanese dog breed) *Kabosu* that went viral in 2010.

Dogecoin initially had no real value (some think it still doesn't). In fact, the crypto-community shunned it for many years, treating it as the unwanted black sheep of the Bitcoin and Litecoin family. Major crypto exchanges like Binance.US and Coinbase were reluctant to list it out of fear of tarnishing their prestige. It was an orphan coin, forgotten.

But then, in 2021, Elon embarked on a Dogecoin campaign that mobilized people worldwide to throw their support behind DOGE. In January of that year, he tweeted a magazine cover titled *DOGUE* with a picture of what looked like an Italian Greyhound modeling a fancy sweater. The magazine was a play on the fashion magazine *VOGUE*, but crypto Twitter interpreted it as a call to action. By the end of that day, DOGE had risen almost 700%.

The tweet caught the attention of the subreddit crypto group *SatoshiStreetBets* (a crypto version of WallStreetBets—the subreddit group whose viral posts caused Game Stop and AMC stocks to surge hundreds of percent in January 2021). They rallied behind Elon and started a side-campaign prompting members to buy DOGE.

From that point, Elon sent several strings of witty (and some not so witty) tweets that propelled DOGE prices and rallied a global gang of Dogecoin supporters known as the *DOGE Army*.

Within a few months after Elon started his Dogecoin campaign (December 31, 2020 to May 2021), DOGE price rose 14,000%, taking it from one of the lowest ranking cryptocurrencies by market cap to the fourth-largest cryptocurrency by market cap, beating popular contenders like Tether, Litecoin, and XRP.[86]

Some People Were Not Cool with This

Elon's Twitter behavior with Dogecoin was even more controversial than with Bitcoin's.

Why was that? Remember what I said in the last section. Dogecoin was created as a joke; it was not made to bring any real value to the world like Bitcoin. Critics inside and outside the crypto-community worried that armature investors (the majority invested in Dogecoin) were not buying DOGE because they thought it had real utility, but because Elon was endorsing it and they wanted to make money. The collective concerns were that Dogecoin's meteoric rise and subsequent fall would devastate investors.

Nevertheless, Elon did not let the criticism stop him. Elon continued to be Elon. And right when you thought he'd hit his limit on DOGE obsession, he leveled up and did something unusual.

Live from New York, the DOGE Debacle

Do you watch Saturday Night Live (SNL)? Or maybe you've seen some of their more controversial sketches make the morning paper or evening news? One great thing about the show is its ability to combine some of the hottest music artists with some of the most unanticipated hosts and put them on one stage for one hilarious night. That is why I nearly jumped out of my shoes when I heard Elon Musk was to host the May 8, 2021 episode of SNL with Miley

Cyrus as the musical guest.

Why was I excited? Of course, I wanted to see Elon, an unpredictable, multi-billionaire business bad boy, act like a fool on stage, but I also knew what his appearance on SNL would mean for Dogecoin prices. SNL captures the eyes and ears of millions of people every episode, and their sketches capture billions of views on YouTube. Without a doubt, SNL is a vehicle through which hosts, musical guests, or even the SNL team can send messages to the world. The DOGE Army hoped Elon's statement would be in the form of investment advice, something in the neighborhood of *"Go buy Dogecoin!"*

The episode went live that Saturday night. Elon was left out of the show's cold open (the skit right before the host goes on stage to do their monologue). Immediately Dogecoin's price began to fall, a sign that DOGE traders got nervous about the episode's prospects for Dogecoin.

After the cold open, Elon jumped on stage to do his monologue. This was his opportunity to tell the world that Dogecoin will one day take over the financial system and that they should go buy some, but no dice; he went the entire monologue without mentioning Dogecoin. Dogecoin's price fell further, faster.

However, as long as there was time left in the episode, there was hope. I'm certain DOGE traders had their thumbs hovering over the "buy" button on an order of DOGE just waiting for Elon to say something inspiring about it.

The moment finally came during *Weekend Update* (a segment poking fun at the news). News anchors Michael Che and Colin Jost repeatedly asked Elon (playing a cryptocurrency expert named *Lloyd Ostertag*) what Dogecoin was. After giving five or six *right* answers,

Lloyd laughed, shrugged his shoulders and said, *It's a hustle*. And that was the end of it. Folks were selling Dogecoin so fast that it broke the Robinhood trading app.

By the next day, Dogecoin had lost almost a third of its market value. The coin's performance numbed the DOGE Army. Top news agencies like CoinDesk, Decrypt, CNBC, CNN, Yahoo!, Rolling Stone, and everyone else you can think of pointed out Dogecoin's embarrassing price drop during and after the SNL episode.

Elon attempted damage control after the SNL episode, stating he was working with Dogecoin developers to improve the network[87] and that he hadn't (and wouldn't) sell any of his DOGE. But still, it failed to return to its highs before his SNL appearance.

Market Manipulation?

Not everyone was excited about Elon's venture into cryptocurrency. Tweeters, Redditers, and even investment professionals accused Elon of manipulating the crypto market. The consensus was that the correlation between some of Elon's tweets and bitcoin and DOGE price movements was not a coincidence and that he purposely drove up (or down) prices to profit financially.

Another argument that supported manipulation allegations was the question of why Elon supported (and purchased) bitcoin in the first place. He's dedicated most of his career to making the world a cleaner place to live. Some critics had a hard time believing that he did not know about Bitcoin mining's effects on the environment before he invested in it, unless his initial goal was to inflate its price and then sell.

But Elon's influence over the crypto market (at least with his tweets) dwindled as crypto entered a bear market in Q4 of 2021. Concerning Dogecoin, some say that his tweets had no lasting effect on Dogecoin's price. For example, Elon posted three tweets about DOGE from May to December 2021, and with each one, DOGE's price spiked slightly, but quickly reversed and continued to trend down with the rest of the bear market. What could have triggered this? Either the crypto-community woke up and realized that they should not allow themselves to be swayed by the sentiments of one person, or the market volatility associated with Elon's tweets was too emotionally taxing and investors no longer wanted to deal with the stress. (Or it could have been for both reasons.)

Who Is to Blame?

So, should Elon be blamed for the Bitcoin crash and DOGE crash of 2021? Well, it depends on which crash is in question. If you were trading crypto then, you'll remember that the crypto market encountered several heart-stopping crashes that wiped out gains from millions of portfolios, including mine. Bitcoin crashes are nothing new, in fact, they're a tradition. But, as I illustrated above, some investors blame Elon for the crashes that corresponded to his tweets. Whether or not he purposely did it to manipulate the crypto market has yet to be determined.

If there is one thing that the Elon Effect has uncovered, it is the philosophical divide between old school crypto investors and new school crypto investors. The sentiment from some old-school crypto investors is that no one in the crypto-community should let one man— especially someone who's barely dipped a toe in the crypto market— control their trading behaviors because it goes against the whole ethos of Bitcoin and decentralization. On the other hand, some new school crypto investors only trade cryptocurrency to make money and do not prescribe cryptocurrency's "power for the people" spirit.

Regardless of whose fault it was, I think one thing we can say for sure is that Elon had an unprecedented effect on the crypto market. Judging from his recent moves in the crypto space, something tells me that we haven't seen the last of Mr. Elon Musk, the Dogefather.

CHAPTER FOUR

RUG PULLS, BAG HOLDERS, AND PUMP-AND-DUMPS

Are Meme Coin Economics Casting Shade on the Crypto Industry?

The meme coins craze in 2021 has made people millions—and some, billions! Career investors, factory workers, high school students, college dropouts, and people from all walks of life invested a few thousand dollars into a meme coin one day, only to wake up the next day learning that they've become filthy rich.

How awesome of feeling that must have been!

Next to non-fungible tokens (NFT), meme coins are the most popular sector in cryptocurrency. Dogecoin, Shiba Inu, and Dogelon Mars, which were created for nothing more than to play around with, dominated headlines on CNBC, Yahoo! Finance, Motley Fool, Bloomberg, CNN, and Fox News. Their fun designs, cute logos, and capacity to increase thousands of percent in price in only a few hours has the world jumping in to get a piece.

With their success in the crypto market and mainstream notoriety, there's no question that the meme economy has become a coveted sector in the crypto industry. Today, there are thousands of meme coins in the crypto market.

But though people enjoy the frivolousness of meme coins, some people believe they are so frivolous that they attract scammers, immature traders, underdeveloped projects, and illegitimacy to the crypto market.

The question now is whether the meme coin economy is negatively affecting the crypto industry. There's been a long debate by leaders, experts, and traders in the crypto-community about this question, and the jury is still out. Both sides make exciting and compelling arguments for and against meme coins, so for us to respond appropriately, I will break down both sides of the argument and conclude with their final answer.

Get ready to have some fun, but first, let's take a quick look into what meme coins are.

What Planet Did Meme Coins Come From?

Shockingly, right here from Earth!

Meme coins are not the same as bitcoin, ether, or DOT. Their instrumental value is not based on a solid project plan or sound fundamentals but on wild speculations, endorsements from irrelevant celebrities, and greed.

I assume that most of you already know what an internet meme is since they've been practically steering humor in pop culture for the last fifteen-plus years. But for good measure, let's recap.

A meme is an internet image that usually exhibits a funny picture with a cleverly-worded statement poking fun at a cultural symbol, social idea, or current event. They are meant to be funny, like a joke.

Meme coins are cryptocurrencies inspired by internet memes. They were created for crypto traders to own and trade for fun—there were no real-world use cases built into them like Bitcoin and Ethereum.[88]

For example, in the previous chapter, you read about Dogecoin, the first meme coin. Dogecoin was branded with the image Kabosu, the side-eye-glancing Shiba Inu from the 2010 internet meme. The word *Doge* originated from a 2005 episode of the puppet web series *Homestar Runner* where the star, Homestar Runner, annoyingly said to his coworker, Strong Bad, *"What is up my dog?!...You crack me up! That's why you're my d-o-g-e (he spells)."*[89] The whole thing was a joke; Dogecoin did not even have a development team backing it at the time.

Developers capitalized on the meme coin craze following Dogecoin's explosion in 2021. Every other day a new animal-related meme coin emerged, like Kishu Inu, Hoge Finance, Dogelon Mars, and of course, the *"Doge Killer,"* Shiba Inu. And since people were making money, more investors jumped into the space and new meme coin projects.

Back to Both Arguments

Okay, now that you have a good idea of what meme coins are and where they came from, let's explore what the groups that are for and those that are against meme coins feel about their influence on the crypto industry.

Group A – Meme Coins Are GOOD for the Crypto Industry

Here are some of the primary reasons meme coin enthusiasts are advocating for meme coins:

Brings New Investors into the Market. The Dogecoin and Shiba Inu craze of 2021 stands out as one of the most significant events in crypto history. This was primarily due to the absurdity at which meme coins rose from the lowest on the crypto ladder to one of the largest by market cap, but it also exposed the world of investing to people who would not have otherwise paid attention.

This is especially true with young teenagers, young adults, and college students. For many of them, Dogecoin was the first cryptocurrency they bought. Some of them discovered other cryptocurrencies and became crypto traders. Likewise, crypto trading probably gave many a worthwhile after-school hobby and a possible future career to aspire to (unlike my childhood hobby which involved Street Fighter 2 and the original Mortal Kombat for SNES).

Forms Thriving Communities. If there is one thing fascinating about meme coins, it's their ability to assemble communities of young supporters from all over the world. Tons of Reddit, Twitter, Discord, and Telegram posts formed around the meme coin economy to discuss best trading practices, impending meme coins, scam warnings, best wallets, airdrops, and other topics. Whether they knew it or not, their participation helped them develop or exercise skills that will benefit them as adults.

Visit one of the Reddit groups and see what I mean. Some (like, *some*) debate why or why not to invest in a coin, present clear arguments to buy one coin over another, use logical reasoning, make sound judgments, interpret meanings, conduct research, think through problems, and propose solutions.

These communities could also be good for friendships. A 2015 study by the Pew Research center found that 64% of teens ages 13 to 17 have made a new friend on social media, with 69% of them indicating that they have made more than five.[90] I can imagine these online communities giving young people, especially introverts, a place where they can form bonds with other like-minded people.

Teaches People How to Handle Loss. Failure is life's ultimate lesson. Whether you experience your own failure or heed the failures of others, it teaches you how to hone your skills so that you don't make the same mistake. Experiencing failure hits a little harder when money is involved. Meme coins are more volatile than other cryptocurrencies, so they can be detrimental to one's crypto portfolio if not traded carefully. And though it's praiseworthy that meme coins attract young, first-time investors to the crypto space, unfortunately, most of them are inexperienced and make copious investment mistakes that cost them hundreds if not thousands of dollars. For young teenagers, that's a lot of money.

They tend to make bad investments because they base investment decisions on emotions and speculation, which opens them up to scams and bad investment advice. Some would think of this as an unfortunate outcome—and it is—but looking at it from a different perspective, it's a valuable experience. Let's face it, it's best to lose money in your younger years because you will have time to correct your approach and rebuild later. Many early victims of trading pitfalls have later evolved into savvy investors.

Group B – Meme Coins Are BAD for the Crypto Industry

With all the value meme coins bring to the crypto industry, why do some people give them a bad rap? Well, here are some of the primary reasons that meme coin haters want meme coins gone for good (DISCLAIMER: Their reasons are way more robust than those of the last group.)

They Don't Trust Them. Some investors don't trust meme coin developers because many of them lack technological (or creative) skills and create projects that lack appropriate software, direction, creativity, and a functioning website.

Attracts Lame Investors. Some investors do not like the quality of investors that meme coins attract into the market—mostly young, inexperienced investors who operate based on emotion rather than on sound principles. They feel that immature investors throw off the crypto market for everyone. Third and most important, investors are uncomfortable with how susceptible the meme coin economy is to scams like rug pulls and pump-and-dumps that leave well-meaning investors rekt (victimized).

Scams, Scams, Scams. Some investors believe that the meme coin economy attracts a buku of scams. Let's first look at bag holding, then explore a few of most common meme coin scams:

Bag Holding – You Feel Like an Idiot

I will be referencing bag holding throughout the rest of this chapter, so it's probably best to tell you what it means. A bag holder is an investor who invests in a particular crypto they're confident will turn a profit but refuses to sell when the price starts to tank.[91] They call it *bag holding* because, while other investors who acknowledged the price declines sold their cryptos and made off with profits, you're left standing there with the *I-didn't-sell-my-stuff-in-time* face, holding the bag of worthless cryptos, feeling stupid. I've been there. It's not a pleasant feeling.

Bag holding itself is not a scam, but it often results from scams. It goes hand-in-hand with rug pulls and P&D schemes. You'll see what I mean in a minute.

Rug Pullers – The Squid Game Project

Rug pulls are probably one of the most common scams in crypto, and they are rampant in meme coins. Rug pulls occur when new projects are launched into the crypto market. This is generally how it unfolds: Someone launches a new meme coin into the space—investors buy the coin expecting a return on their investment—then suddenly, the project developer closes the project, sells all the coins, and disappears with all the cash. Investors are left holding the bag. Classic rug pull.

A great example of this happened in 2021 with the Squid Game project that surged 230,000% in a week to $2,850, only to fall 99% in less than a second to 0.5 cents.[92]

The Squid Game project was a play-to-earn blockchain game inspired by the uber popular Korean Netflix series *Squid Game*. Squid Game developers launched the project in late October and it was an instant success.

To play, investors paid an entrance fee in Squid Game tokens which unlocked a series of six games that players had to complete to win the prize of 90% of the entrance fees (the other 10% would go to developers). Those who completed the first game advanced to the second game—likewise, those who completed the second game advanced to the third game, and so on. The last survivor in the sixth game would win 90% of the entrance fees.

Influencers on Twitter, Reddit, YouTube, and every other social media platform raved about the game so much that other investors joined the action out of FOMO (fear of missing out). The token price skyrocketed from barely a penny to $2,850.

Oddly enough, investors reported that they could not cash out on their Squid Game tokens on the project's webpage (It was likely due to an anti-dump software program that ran on the site, but who knows?)

Sadly, on November 1, one week after the Squid Game project went live, their token price fell to zero. Their website and all of its social media accounts vanished. Developers cashed out on all the tokens and made off with $3,387,387.59 of the investors' money.[93] Rug pull!

Pump-and-Dumps—the Worst Feeling Ever

Pump-and-Dumps (P&D) are probably one of the most widely used scams in the crypto industry. They are become something of a crypto custom.

A P&D is a type of stock market manipulation whereby a company (or a person or group of people who hold shares in the company) artificially pump up the company's stock price by making false and/or misleading positive statements in order to sell the stock at the higher price later, leaving all the investors that bought into the stock holding the bag.

People caution against trading meme coins because they are more susceptible to P&Ds than any other genre of coins. This is mainly because of their low price. If you've been trading crypto (or even equity stocks), you've probably been caught in a pump-and-dump or two.

I was caught in a P&D scheme when I started trading crypto. I considered myself a pretty savvy investor since I had already traded stocks and had done so for years. However, the amount of information on the crypto market and crypto trading dos and don'ts weren't as robust back then as they are today, so I didn't have much to go on but trial and error. Well, as many crypto traders did (and do), I rushed in with a bag full of cash and bet it all on a particular crypto, and then boom! Not even two days later I lost almost everything. (I won't name names, but that crypto is still around today and doing very well). It sucked. So, if you've ever fallen for one of these, I feel your pain!

With respect to the stock market, pump-and-dumps are illegal under the Securities Act of 1933, the Securities Exchange Act of 1934, and several sections of Title 18 of the U.S. Code.[94] Violators are punishable by fine or jail time. However, there are no such regulations against pump-and-dumps in cryptocurrency, so they are rampant.

Pump-and-dumps schemes have become an organized strategy by a group of investors rather than a scam. There are several techniques that P&D groups use to pull one off, but what they will typically do is select a specific coin (especially meme coins) to pump, choose a date for when the pump will begin, start a massive baiting campaign on Twitter, Discord, Reddit, Telegram, etcetera about the coin, and use keywords and phrases that evoke an emotional response like "it's about to explode," "buy now," "just bought [number] or [a coin]."

And just when the price of the coin rises to a desired level, the group will sell and take the profits, leaving the people that fell for the bait holding the bag. Easy money.

The Psychology of Meme Coins

I know you all are waiting for the answer to the million-dollar question, but I would be remiss if I did not explain some of the psychology behind meme coins, particularly what draws people in and keeps them sticking around. Some of the psychological effects are spurred by market sentiment, others are created by crypto developers who use techniques based on marketing psychology, behavioral psychology, and prospect theory in their project designs and advertisements. So, bear with me for a few more paragraphs, I promise it will be worth it.

Greed. Next to love and fear, greed is one of the fiercest motivators. There is a lot of money to be made in meme coins, and people prove it to us every day. In October 2021, there was a story in online cryptocurrency news outlet, Cointelegraph, about a guy who turned $3,400 to $1.5 billion in less than a year.[95] The Independent reported a story of a 35-year-old factory worker in the U.K. who turned 6000 British pounds (GBP) into millions.

Some people read these stories and take unnecessary financial risks to obtain that kind of wealth, even if it means being scammed. There are people who lose both their primary and emergency savings in risky meme coin projects. Some risk their car payment or rent money. Greed has a way of blinding us to bad investments. Investors, especially inexperienced ones, fail to approach projects with common sense because there is often a lot of easy money to be made. This, combined with the hype cycle spurred by communities of crypto "influencers" on YouTube, Twitter, and Reddit baiting them to buy the newest, hottest coin, creates an "invest now, ask later" mentality, which will ultimately come back to bite them later.

Let's look at how meme coin design and marketing tactics coincide with psychological principles.

Perceived Value. Question: Which of these two values looks more enticing? 0.000032 bitcoin or 601,684 Shiba Inu coins? To the untrained eye, or in this case, the inexperienced investor, they will probably say, *"Hey, owning 601,684 Shiba Inu coins looks like a better payoff than owing 0.000032 bitcoin."* Why is that? It's because they *perceive* that more coins equal more profits. In market psychology, this is referred to as *perceived value*. Businesses use several pillars of perceived value to sell products, but perception is not always the same as reality. 601,684 Shiba Inu coins may look more valuable than 0.000032 bitcoin, but both quantities are worth $10 at the time of this writing.

Perceived Value of Price. Have you ever wondered why meme coins are so cheap compared to other coins? It is because they are minted by the quadrillions.

Bitcoin has a total supply of 21,000,000 coins.[96] This means that, as new bitcoin are minted, their supply will decrease, making them more valuable to the buyer and thus, more expensive (this is the Psychology of Scarcity). On the other hand, Shiba Inu has a total supply of about 999,991,456,178,974 coins.[97] (Yes, that's almost one quadrillion!) That's 47,618,640 times bitcoin supply and nearly 129 times Earth's population. There are obviously more Shiba Inu to go around, so its price will be significantly lower—fractions of a penny.

Meme coin developers are quite cunning. They create quadrillions of coins because, visually, owning millions of coins looks enticing to the buyer. Inexperienced investors may not understand how a project's market cap and total coin supply factor into the coin's overall price and how high it will rise. Therefore, when they see the opportunity to buy 35,000,000 coins for only $500, they capitalize on it, believing that one day the coin's price will reach twenty-five cents, fifty cents, or even a dollar and make them filthy rich.

Psychology of Social Proof. Have you ever searched YouTube for information on a new coin and pulled up a ton of those annoying video images of a guy with his mouth wide open in shock and dollar sign emojis circling around him? Those guys are called *influencers,* and their job is to get you to watch their videos and buy the coin they're advertising.

The idea behind social proof is that people judge whether to do something based on whether others are doing it. Companies use social proof to legitimize their products and make more sales. Nike uses famous athletes to advertise their products, FTX once used Matt Damon (good move), and meme coin developers sometimes use social media influencers that no one's heard of. But their trick is to use a vast array of influencers from several social media platforms to promote their coin, giving the illusion that the "experts" vouch for it.

Think about it, if enough people are saying the right things about a particular coin, there is a better chance that people will buy it.

Loss Aversion (AKA FOMO). People hate losing things, especially money. And though people hate losing money in the crypto market, they hate missing out on a lucrative investment even more. Loss aversion is a term by psychologists Daniel Kahneman and Amos Tversky[98] that says people would rather prevent the loss of something than gain the same thing. For example, the pain of losing $50 in the market outweighs the pleasure of earning $50. Odd, right? But I'm sure you can testify from personal experience that they are dead on.

The interesting thing about loss aversion, however, is that you do not necessarily have to own the thing you fear losing, you could fear losing the chance to be a part of it. That is why FOMO is such a strong emotion, and businesses use FOMO tactics in their marketing to sell products.

Great examples of this are the buzz phrases that retail stores use to create a sense of urgency in customers:

- *"Only five left in stock! Order now!"*
- *"Four Shoppers Added This To Their Cart."*
- *"Get them while supplies last."*
- *"One Day Sale, Don't Miss Out!*

You all know what I'm talking about.

Circling back to the social proof concept, FOMO is a tactic that social media influencers use to prompt investors to jump into the action. They use buzz phrases like:

- *"This Coin Is About to Blow!!"*
- *"[coin] 100x Potential!"*
- *"Become a Millionaire with [coin]"*

Now, FOMO happens with all cryptocurrencies, but because meme coin prices are cheap and can soar sky-high in a short period, the FOMO is much more intense. So, as the loss aversion principle would have it, investors naturally do not want to miss out on making money, and they jump into an investment without doing proper research.

Color Psychology. Did you know that color can influence our purchasing decisions? Yes, it's a fact laid out in *color psychology*. In color psychology, businesses strategically use colors that align with the emotions they want to provoke.[99]

For example, *yellow* is associated with sunshine, fun, happiness, and low prices. Retailers usually mark sale and clearance items with a yellow tag because customers are naturally happy when paying less for things they want. *Red* is associated with excitement and urgency. Netflix, CNN, Target, Coca Cola—these brands evoke excitement (and some urgency) when people see them. The enthusiasm will naturally draw customers to them and their products.

Most crypto projects use color psychology in their branding, especially meme coin projects. People are visual creatures. They buy meme coins because they evoke excitement, fun, happiness, and joy.

That is why most meme coins have logos that feature cute, colorful cartoon animals that would draw the attention of anyone with a soul. Likewise, the project's website is usually drenched in bright, excitement-provoking colors to draw in the viewer.

What's the Answer?

Okay, now you've heard the arguments for and against meme coins, the question remains: Are meme coins bad for the crypto industry?

Answer: Not necessarily.

This is probably not the answer you expected, but considering the facts in this chapter, meme coins' benefits and disadvantages do not permit a conclusive answer.

However, I believe there is a way to reduce the negative impact of scams on the crypto market and reduce the frequency of investors becoming victims of lousy meme coin projects. No, it's not with regulation but with education. If history teaches us anything, people will use innovations to do bad things. This will not go away even with regulation. But financial literacy can help. If investors are taught how to spot scams before becoming victims, then negative sentiment around meme coins will decrease.

I know, it's a simple solution, but a necessary one because despite all the meme coin benefits to the crypto market, no one likes to lose money.

Meme coins have communities of developers, investors, enthusiasts, and celebrities backing them, so I don't think they are going anywhere anytime soon. They did well in drawing new investors to the market, promoting incredible communities, driving innovation, and teaching valuable life lessons.

However, if we want to block the negative shade they cast over the crypto market, we (the crypto-community) have to do a better job of weeding out scams. Let's spread some education. I've included a step-by-step crypto research guide in the back of the book. Use it. It might seem like a drag, but the more research you do upfront on a project, the less disappointed you will experience later. Make it a goal to scam-proof your meme coin portfolio, it will thank you later.

CHAPTER FIVE

THE ETHEREUM EXPERIENCE
How Gas Fees Killed Our Vibe

I f you made at least one Ethereum transaction in 2021, then I guarantee you've had the Ethereum experience, too—you just didn't know that you had it or that there was a name for it.

When you first hear *Ethereum Experience,* you probably think of something magical like the *Meta Quest 2 Virtual Reality Experience* at Disney World or the *IMAX Experience* at the movies. Well, you shouldn't. The Ethereum Experience I'm talking about started around Q2 2021, and there was nothing magical about it. In fact, the Ethereum experience forced congregations of crypto traders and developers away from Ethereum to pursue other blockchains. It turned other traders into Ethereum-haters completely.

Like you, I've had the Ethereum experience several times over, and I still struggle with it.

The Ethereum experience goes something like this:
It's a typical day; you feel good about the crypto market and want to make some trades. You log onto your preferred crypto exchange or wallet to buy/send/swap ether or Ethereum-based tokens. You choose the tokens that you want to trade and begin the transaction. Everything is going well until you get to the page showing the transaction's estimated gas fees. You look at the price of the gas fee and in sticker-shock fashion, you do a double-take because there is no way gas fees are THAT high. You spend about five minutes fiddling with the exchange options wondering if you pushed a wrong button or if there is a glitch in the system screwing up the gas prices. At the end of those five minutes, you realize that the crypto exchange is working fine and there's no glitch. You begin to worry. You sit for another five minutes trying to make sense of how the Ethereum network allowed gas fees to be THAT high for one trade. At this point, you're upset and trying to decide whether the trade is even worth the money you'll shell out for it.

At the end of those five minutes you're going to do one of two things: 1) You will either bite the bullet and pay the gas fee and pray that you get a decent return on your investment, or 2) You will say something like, *"I'm not F***ing paying this fee for a trade,"* throw your phone or laptop against the wall, and storm out of whichever room you're in. No matter which path you choose, you almost certainly will spend the rest of the day pissed off at everyone and everything and just want to punch something.

Congratulations, ladies and gentlemen, you've just had the Ethereum experience.

We're going to dissect this experience a bit more in this chapter and explore what Ethereum is, why gas fees are so high, how Ethereum fell victim to its competitors, and what the Ethereum Foundation is doing to win back user confidence.

Why Do People Love Blockchain?

Let's take a look back at why people fell in love with blockchain. People fell in love with blockchain because of its practical benefits. It has been a savior to families who could not afford to pay high transaction fees for cross-border payments and a savior to traders who were fed up paying overpriced broker fees on common stock trades.

Have you ever wondered why stock brokerages charge traders $7–$12 to buy and sell stocks? Or why commercial banks charge $25–$30 to wire money to another account? If nothing comes to mind, then you're correct because there is no reason! At least not a logical one. Banks and money transmitters just want to see how much money they can reasonably (and legally) squeeze out of you because there are no better alternatives available. The traditional financial system doesn't make sense and transaction fees are just too dang high! So, you can imagine people's relief when they discovered blockchain and cryptocurrency. They were no longer slaves to centralized institutions' overpriced transaction fees.

Blockchain gave me—and slews of people—hope for an affordable way to do money. We were excited about blockchain and cryptocurrency's hope for the world.

But then, in 2021, Ethereum gas fees happened.

Ethereum

Ethereum is the second largest cryptocurrency project by market cap.[100] It's the premier blockchain that underpins some of the most popular decentralized finance (DeFi) and NFT applications. People in the crypto community recognize Ethereum for its OG status, versatility, and multiple real-world use cases. And as the second most popular blockchain next to Bitcoin, it is commonly known as the blockchain of choice for developers looking to build solid crypto projects. Regrettably, however, Ethereum is also known for something else which has shown to be quite unpopular with its users—its flagrantly high gas fees.

If you were to ask any trader in 2021 what they disliked most about Ethereum, their answer would undoubtedly be its high gas fees. Users must pay a gas fee when they make transactions on Ethereum. *Gas* is like the fuel that moves the transaction through the network until it is complete. The amount of gas required for each transaction depends on the complexity of the transaction. For example, you might pay a small gas fee for your standard transaction but pay a higher one for something like a token swap.

Ethereum was not originally designed to give users heart attacks over the price of gas fees but to create a more versatile blockchain than Bitcoin. Ethereum was founded in 2014 by super-genius coding guru Vitalik Buterin. The project was officially launched in 2015 after a 42-day ether coin sale the previous year. Ethereum introduced a built-in, fully-fledged Turing-complete programming language that could be used to create algorithm-based "contracts." These contracts can be coded to execute any command automatically (we'll address this further, shortly). It was a developer's dream. And unlike Bitcoin, whose programming language limits its functionalities, Ethereum's Solidity-based programing language promotes near-infinite processes.

Ethereum Grew on Me

I encountered Ethereum in 2018, and to me, it didn't seem any different than the other projects out there, only that it checked all the boxes for me to start trading Ether—it was a blockchain, it had its own cryptocurrency, and it was available on all the leading crypto exchanges. Honestly, I didn't bother to learn how Ethereum's network operated nor cared about how it differed from other blockchains. I was a crypto newbie and naïve; I was only concerned about making money from it and didn't care about the technicals.

But people on YouTube and Twitter talked about Ethereum nearly non-stop. Many said it was better than any other blockchain, even Bitcoin. That shocked me. Bitcoin was the mecca, the motherland, the father of cryptocurrencies. So, when people said that Ethereum was better than Bitcoin, it made me curious and I started digging into what Ethereum was all about.

I read every article and watched every video about Ethereum, and I noticed one constant: they emphasized this thing called a *smart contract*. *Smart contract* this and *smart contract* that. The term came up so much that it started to annoy me, so I did the smart thing and investigated this *smart contract* thing everyone was talking about.

"A smart contract is a body of computer codes that execute commands on a blockchain. You can program the agreed-upon terms of any transaction into a smart contract and it will automatically execute the transaction once the agreed-upon terms are met. There are no brokers, no banks, no mediators, or third-party payment processors. There are only the buyer, the seller, and the smart contract."

Smart contracts changed the game forever. They enabled developers to build cool decentralized applications like DeFi, oracles, decentralized exchanges (DEX), NFT platforms, prediction markets projects, and more. I didn't understand its impact right away, but as more developers opted to build on Ethereum, it became clear that Ethereum was going to change the way people did money and create opportunities for broader financial inclusion.

The Rise Of DeFi and the Bad Blood that Followed

DeFi is arguably the most innovative breakthrough in cryptocurrency since Ethereum. For those who don't know, DeFi is a peer-to-peer financial decentralized application that runs on a smart contract. It offers various conventional financial services like loan services (borrowing and lending), interest accounts, asset management services, DEXs, derivatives marketplaces, and insurance.

DeFi had the potential to bridge the wealth gap between poor, underserved communities, and the wealthy. It needed no bankers, brokers, loan applications or credit checks, nor did it discriminate against people based on race, gender, or social class. People only needed a computer, an internet connection, an open mind, and some faith to get started.

DeFi's explosion in late 2019–2021 confirmed the market's hunger for alternative financial services, with DeFi lending platforms like Maker, Aave, and Compound leading the way. Between October 2019 and December 2021, the total value locked (TVL) in DeFi protocols increased a staggering 51,000% from $463 million to $237 billion. In 2021 alone, the TVL increased nearly 1,295% from $17 billion in January to $237 billion in December, according to Defi Lama data.[101] We've never seen these kinds of numbers in the traditional stock market.

Investors were making millions, and miners, billions. Much of De-Fi's wealth explosion can be attributed to a nifty DeFi feature called a *liquidity pool*.

Liquidity pools are the backbone of DeFi protocols and their most used feature. Think of a liquidity pool as an interest-bearing savings account. When you deposit money into a savings account at your local bank, you essentially provide liquidity to the bank. The bank takes the money you deposited and loans it to borrowers. The borrower pays the bank interest on the loan, and the bank in turn pays you interest on your deposit.

A liquidity pool operates the same way. When you deposit tokens into a DeFi liquidity pool, other users can borrow them and sell them on a crypto exchange for cash. In return, the DeFi protocol will pay you interest on the deposited tokens.[102]

But here is the difference between the two: While banks typically pay between 1% and 2% APY on your savings deposit, most DeFi liquidity pools pay between 10% and 30% APY. These returns were unheard of in the traditional banking industry, which explains why many people, rich and poor, ditched conventional bank accounts for DeFi.

Nevertheless, despite DeFi's popularity with crypto investors and its limitless capacity to help the world do money better, DeFi's explosion had catastrophic effects on Ethereum gas prices, which led to some major setbacks for Ethereum and its users.

The Gas Fee System Is Flawed

Unfortunately, Ethereum's gas fee protocol is flawed and can be detrimental to your average Ethereum user. Here's why:

Ethereum transactions operate on an auction-like system:

- Gas is the currency

- Transactions are the auction items

- Ethereum miners are the auctioneers

Because Ethereum block sizes are only 12.5 megabytes, and there are only so many miners on the network processing transactions, users have to incentivize the miners to process their transaction(s) first by offering a higher gas bid.[103]

Think of it this way; when network activity is low, there is less competition, and miners can process transactions indiscriminately at a lower gas fee. But when network activity is high—that is, when miners get overwhelmed by the high number of transactions waiting to be processed—they will process the transaction with the highest gas fees first, regardless of the order in which the transactions were made.

Why do they do that? Because Ethereum miners are paid in gas fees. They will naturally go after the transactions with the highest payout.

Wouldn't you?

Vitalik certainly meant well when he originally designed the reward system, but as it turns out, it presents a problem: Since Ethereum transactions operate like an auction, gas fees will continually trend higher when the network is congested with transactions. This feature will eventually freeze out average users because they cannot afford the gas fee required to make transactions.

But we've seen this play out before. The Ethereum network saw its first major challenge during the Cryptokitties crazy of 2017. These NFTs were in such high demand that transactions nearly brought the network to a screeching halt. Miners could not keep up. The network congestion did not last very long, but it signaled a bigger issue with the network's composition and fundamentals.

It's important to mention that the gas limit for a transaction remains the same regardless of how the gas price fluctuates. To put this into perspective, the gas tank in your car will require the same amount of fuel to fill, irrespective of how long you've had the car. However, the price of fuel might fluctuate based on the price of oil at the time. So, if in 2003 you drove a mid-size sedan in the Midwest, you probably paid around $22 to fill the 12-gallon tank. If you were still driving the same sedan in 2022, you're probably paying around $57 to fill the tank.

To draw the parallel with Ethereum, if the gas limit to transact one Ether is 1,000, it will always be 1,000. However, the gas price will change based on network activity. But unlike buying fuel for your car where you must pay the price displayed at the pump, Ethereum users can set their gas price based on how fast they want the transaction to be processed (This is the auction system we talked about above).

The Gas Crisis of 2020-2021

The initial spike in gas prices in 2020 and 2021 resulted from the high demand for DeFi services. Gas fees more than tripled between October 2020 and March 2021.[104] I can attest that the sudden spike in token swap fees blindsided traders. It felt as if one day everyone was happily swapping tokens in their Metamask wallets and the next day passing out from the sticker shock of the gas fees. The spike hit us where it hurt, in our pockets. As gas fees climbed, so did animosity toward the Ethereum network.

The crypto community took to Twitter, Reddit, and YouTube to voice their complaints. Traders testified to paying upward of one-third of their total transaction amount in gas fees. Some traders couldn't afford to buy Ethereum-based tokens because the cost to buy the token was higher than the full purchase price. A few members in the community suggested a boycott of Ethereum-based DEXs (which would be futile—but still…).

It was mayhem. Rising gas fees became every major crypto news article and podcast headline. Traders sought out every option to mitigate the gas fees. I remember waking up at the butt-crack of dawn to make trades because the gas fees were slightly less in the early morning than during the afternoon and evening.

Some Could Not Say No to Ethereum

Hating Ethereum is difficult. Despite the gas crisis, Ethereum was the premier blockchain of choice for developers to build projects. Don't get me wrong, members of the crypto community, including myself, wanted to stay bitter toward the network, but the projects that developers were building on Ethereum in 2021 were so innovative, revolutionary, and lucrative that it was impossible to say *no* to them.

Tokens for popular projects like Axie Infinity and Shiba Inu at one time only traded on Ethereum-based DEXs like Uniswap or Sushiswap, so one had no choice but to pay the high gas fees to purchase them. I remember ponying up hundreds of dollars on gas fees to buy some of the *must-have* tokens. Thankfully, they've all returned generous profits. However, I'd be lying if I said that I've never been burned by the gas fees or missed out on a good investment opportunity because I could not afford to make the trade.

But Enough Was Enough

Things were getting out of hand and traders were livid. We felt as if we no longer had the same freedoms to trade on Ethereum that we once enjoyed because we literally could not afford to use it. Some of us even had to wait several hours for our trades to finalize, which put us at risk for front-running attacks. Ethereum seemingly caused more financial problems than alleviating them, which of course defeated the purpose of blockchain.

In a sense, Ethereum betrayed the crypto community. The network became dog-eat-dog and exclusive—only the rich and those well off enough to afford to trade could get anything done. People had genuine love for Ethereum. It brought a new meaning to the blockchain and crypto space, but enough was enough. We were done paying 30%–150% in gas fees and sought other, less expensive blockchains to do business with.

Ethereum Killers

Thankfully, there were other blockchain projects ready to scoop up the horde of disgruntled users that Ethereum forced onto the sidelines. The crypto community called these blockchains *Ethereum Killers*. Ethereum killers were high-quality blockchains that, like Ethereum, supported smart contract functionality.

Ethereum killers like Cardano, Binance Chain, Solana, Avalanche, and Harmony took advantage of Ethereum's gas crisis and encouraged new and existing Ethereum developers to build on their networks, promising faster transactions, ease of use, direct bridging to the Ethereum network, and most importantly, *no* gas fees.

Low transaction costs meant users could transact thousands of dollars' worth of crypto for only a few bucks. Fast transaction speeds meant users did not have to wait hours for their transactions to finalize. I remember making $7,000 worth of transactions, including swaps, and only paying a total of $10 in transaction fees. Developers recognized the opportunities in Ethereum killers. The craze brought in hundreds of developers and millions of users.

You could feel the tide shifting. The community felt a sense of relief. Ethereum no longer held the monopoly on innovative decentralized applications. Developers on other blockchains were pushing out projects with just as much real-world value and utility as those on Ethereum. By offering DeFi, NFTs, and staking rewards, Ethereum killers made their own seats at the table.

The numbers speak for themselves. In Q3 of 2021, the number of new users of Binance Smart Chain rose 20% from the previous quarter while its TVL registered $17.78 billion, an 18% growth quarter-over-quarter.[105] Solana soared over 8,000% by Q4 despite suffering denial of service attacks and network outages. Likewise, Avalanche and Terra (before the horrible crash in 2022) saw outstanding gains in Q3 and Q4 due to institutional interest and investments. Even new projects on Ethereum created contracts that allowed people to buy their tokens through other blockchains, bypassing Ethereum gas fees altogether.

Ethereum's Response: Eth2 and EIP 1559

But Ethereum refused to be outdone. The Ethereum Foundation, a non-profit organization dedicated to supporting Ethereum and related technologies, has made tremendous strides in remedying some of the issues wreaking havoc on the network and the crypto community.

Developers understood that if Ethereum was going to meet the growing institutional demand and continue its reign as the world's premier blockchain, something had to be done to make transactions faster. So, in 2021, the Ethereum Foundation proposed a software upgrade to make the Ethereum mainnet more scalable, secure, and sustainable. This new upgrade is known as *Ethereum 2.0.*

In late 2022, Ethereum 2.0., or *Eth2 (better known as, "The Merge")*, replaced Ethereum 1.0's proof-of-work consensus mechanism and implemented the more scalable proof-of-stake consensus mechanism.[106]

At the time that I am writing this, it's too soon to tell if Eth2 is a success as it was recently implemented, it will obviously take some time capture measurable data and work out any kinks, but it should enable the Ethereum network to handle higher transaction demand.

So How Will this Help with Gas Fees?

It won't. At least, not directly. Remember, Eth2 was designed to solve Ethereum's scalability issues, gas fees were not the primary target. However, because Ethereum will be able to handle significantly higher transactions at one time, users will have less need to compete for their transaction to be processed in a timely manner. So, yes, theoretically, gas fees *should* be reduced (but no promises).

Eth2 is a big step in the right direction, but the community also needed assurances from Vitalik and the folks at the Ethereum Foundation that they would tackle the gas problem head-on. Don't get me wrong, users understood that increased tps would make a strong case for Ethereum's future adoption by major institutions and governments, which was fine, especially for long-term Ether holders, but people were hurting and needed a solution that dealt directly with gas fees ASAP.

The Ethereum Foundation decided to double down on their efforts to lower overall transaction fees and win back the community's confidence. *Ethereum Improvement Proposal (EIP) 1559* was part of the long-anticipated London Hard Fork upgrade in August 2021.[107] Without going deep into the technical details, EIP 1559 sought to reduce total transaction fee volatility by increasing Ethereum block sizes and creating a "base fee" for transactions. This essentially made transaction fees more predictable so that users did not have to overpay on gas fees during bidding wars.

But did this happen? Yes and no. The crypto community is divided on whether the upgrade actually improved the network, especially since gas-intensive transactions like NFT minting still drove gas fees higher (this may not be the case for token transaction and swaps).

How this Should Play Out

Ethereum has certainly had better moments. The Ethereum Experience created negative experiences for thousands of traders. Traders felt excluded, betrayed, and taken advantage of.

But the Ethereum Experience did not sour the community's infatuation with blockchain and cryptocurrency. In fact, developers learned from Ethereum's troubles and created user-friendlier blockchains and layer-2 solutions for their users. The crypto community also realized that even game-changing projects like Ethereum will have their hang-ups. And it's okay; we learn from them and do better next time.

People miss Ethereum. I miss Ethereum. I miss trading on Ethereum without dreading gas fees. Traders are ready to put aside the bad blood and return to their first love. Eth2 and EIP 1559 are a significant step in the right direction. If implemented correctly, I believe Ethereum will remain the premier blockchain for decentralized applications and real-world solutions.

CHAPTER SIX

REGULATORS
When You Feel the Hate, Regulate

R egulation—it's complicated. But stick with me here because it's also extremely important. Federal regulation will have a critical role in cryptocurrency's adoption in America and may decide whether the country will be a competitive player in the crypto—and maybe even the FINTEC—industry moving forward.

I've decided to focus solely on America's federal regulatory climate because I have lived in America most of my life and am witnessing the effects of federal enforcement actions against crypto companies, entrepreneurs, and innovators and the regulatory uncertainties affecting the entire crypto space.

This chapter by no means captures the totality of the old, new, and forthcoming regulations and enforcement actions by some of the premier federal agencies in Washington—that will require a book in itself—but I do aim to capture the spirit and motivations of federal regulators' decisions to regulate cryptocurrency and some of the subsequent unintended (or perhaps, intended) consequences.

Crypto Crackdown – the Early Years

I've been following developments in crypto regulation since 2018. I admit that back then, crypto regulation wasn't the most exciting topic in cryptocurrency, but it was one of the more critical topics and one of the most overlooked by the crypto community.

This was mainly because crypto was not receiving the mainstream attention that it is today. Your average trader didn't anticipate federal agencies, specifically the SEC, cracking down on crypto with the fervor they are today. However, the SEC has attempted to quell the crypto craze in the past by taking enforcement actions against alleged *violators*.

Their first enforcement action against crypto took place in July of 2013 when they filed a civil lawsuit against Trendon Shavers and Bitcoin Savings and Trust, alleging that Trendon, operating under the screenname "Pirate40" at the time, was the author of a bitcoin Ponzi scheme that defrauded investors out of more than 700,000 bitcoin.[108] After two months of litigation, the U.S. District Court ruled against Trendon, and he was ordered to pay more than $40 million in disgorgement and prejudgment interest and $150,000 in civil penalties.

A year later, the SEC filed two more lawsuits against crypto companies for the offer and sale of unregistered securities and operating online venues for trading securities (bitcoin).[109] [110]

As the crypto space evolved, enforcement actions ranging from cease-and-desist proceedings to emergency asset freezes gained momentum.

The ICO boom of 2017 triggered a reaction in federal agencies that led to the SEC filing more lawsuits than ever before. Between 2018 and 2020, the SEC brought 85 actions against crypto companies including trading suspensions, litigations, delinquent filings, and administrative proceedings[111] That's seven times the total number of enforcement actions filed during the previous five years. Lawmakers were not silent either. Many of them publicly attributed cryptocurrency exclusively to criminal activities, while others called for it to be abolished entirely from the U.S.

According to a 2021 Cornerstone Research report, between July 1, 2013, and December 31, 2020, the SEC brought 58 actions litigations, 39 administrative proceedings, 20 trading suspension orders, and 10 delinquent filing orders against cryptocurrency issuers, brokers, exchanges, and other service providers.[112]

The Crypto Question – Are They Securities or Are They Something Else?

The crypto boom of 2020–2021 will go down in history as one of the most profound moments in crypto. Bitcoin reached an unbelievable all-time high, institutional investors dropped billions into the space, and crypto became a regular conversation topic in national media and boardrooms across the globe. Investors were happy, entrepreneurs in the space were even happier, and crypto was well on its way to massive mainstream adoption.

Yet, despite all the successes in the crypto space, there has been an undercurrent of anxiety ever-present in the minds of crypto entrepreneurs and investors that emanates from obscurities in the federal regulatory framework for digital asset. Chief among these is the lack of clarity on if/when a digital asset is legally considered a *security*.

The Howey Test

The Securities Act of 1933 provides a list of assets that meet the legal definition of a *security*.[113] These include the usual stocks, bonds, futures, notes, swaps, etcetera. Also on the list is an obscure asset called an *Investment Contract*. Investment contract is reserved for investments that don't fit into the other categories of securities, yet may possess qualities that qualify them as a security.

This is the designation the SEC uses to classify cryptocurrency as a security, sometimes non-discriminately. However, before an investment can be designated as a security under an investment contract, it must meet a set of standards outlined in the Howey Test.[114]

....

The Howey Test

In 1946, the W. J. Howey Company and their subsidiary, Howey-in-the-Hills Service, Inc., were sued by the SEC for the unregistered sale of a security under the Security Act. To provide a little backstory on this case, the W. J. Howey Company owned large acreage of citrus groves in Lake County, Florida and sold plots of the land to interested investors. Investors were offered the option to sign a contract with Howey-in-the-Hills Service, Inc. to maintain the land, cultivate the produce, and sell it on investors' behalf. However, there was a catch, the service contract gave Howey-in-the-Hills complete control of the land, and investors were not allowed to cultivate the land or sell the produce independently. Instead, investors were to share the profits from the produce sales. The SEC sued both W. J. Howey Company and Howey-in-the-Hills Service, Inc., alleging that the land sale and service contract constituted an "investment contract" under the Securities Act and should therefore be considered a security. "Investment contract" however, is subjective in its interpretation. Several kinds of transactions may fall within its purview, so in an effort to clarify an investment contract, the Supreme Court designed what's called the "Howey Test." Under the Howey test, an investment contract is defined as:

```
"a transaction or scheme whereby a person invests
his money in a common enterprise and is led to
expect profits solely from the efforts of the pro-
moter or a third party."
```

A transaction must meet four criteria to be considered an investment contract under the Howey test:

1. An investment of money, 2. in a common enterprise, 3. with the expectation of profit, 4. to be derived from the efforts of others.

The Howey case fit the mold as such: Investors invested money in a common enterprise (citrus groves) and expected to profit from crop sales conducted by a third party (Howey-in-the-Hills, Inc.).

The Supreme Court ruled against Howey et al. and the Howey Test became the standard for securities designations under an investment contract.

....

Crazy stuff, right?

There has been a long debate in Washington on whether current federal securities laws, including the 75-year-old Howey test, could be appropriately applied to an asset class as new and innovative as cryptocurrency. Just looking at it from a common sense perspective, based on its definition alone, the Howey test doesn't always seem like the best determinant of whether something should be considered a security.

For example, the citrus groves owned by W. J. Howey Company were not in themselves a security. In themselves, they can be thought of as property acquired from a sale. However, the process through which the citrus groves were sold and turned a profit for investors classified them as an *investment contract* per the Howey test.

Now, in applying the Howey test to cryptocurrency, the lines get blurred because not all crypto projects raise funds through ICOs or crowdfunding (no investment of money), not everyone expects to turn a profit from coins and tokens (this especially applies to utility tokens), and not all crypto projects rely on the efforts of others to turn a profit. (Think, truly decentralized project. Also, some projects start centrally controlled, but as more investors jump into the project ownership evens out.)

So, with this in mind, can we conclusively say that the Howey test is an appropriate tool to measure whether a cryptocurrency is a security? Well, again, it depends. The Howey test won't apply to every project, so they will have to be reviewed on a case-by-case basis.

Let's Get Together and Talk About It

Another ongoing debate in Washington is the question of which federal regulatory body would regulate cryptocurrency. I thought this was a pretty interesting situation given that cryptocurrency touches almost every aspect of the financial system—from payments to spot markets, to derivatives markets, to insurance, to borrowing and lending—and could fall within the jurisdiction of multiple regulators (SEC, CFTC, OCC, FDIC, or other).

Nonetheless, the SEC has taken the lead on crypto enforcement actions, so they are the ones that crypto entrepreneurs hit up for regulatory clarity. And yet, the only real guidance given to the crypto industry is non-descriptive, primarily derived from SEC letters, reports, settlement action language, and the Framework for "Investment Contract" Analysis of Digital Assets. (This is geared toward ICOs and is extremely vague and confusing. According to the SEC it *is not a rule, regulation, or statement of the Commission, and the Commission has neither approved nor disapproved its content.*)

In the past, crypto companies like Coinbase and many others have attempted to engage regulators about the dos and don'ts of crypto distribution and trading but were met with a lot of silence.

In the meantime, the crypto industry still feels the brunt of the SEC clampdown.

Under the leadership of former SEC chair, Jay Clayton, the SEC filed 56 enforcement actions against crypto companies and entrepreneurs during his three-year tenure. This led some of the most influential shakers in the space to shut down operations:

- In November 2018, the SEC settled a lawsuit against Ether-Delta founder Zachary Coburn for running an unregistered national securities exchange.[115] The SEC cited that almost all of the orders placed through EtherDelta's platform were traded after the SEC issued its 2017 DAO Report, which concluded that certain digital assets, such as DAO tokens, were securities and that platforms that offered trading of these digital asset securities would be subject to the SEC's requirement that exchanges register or operate under an exemption. Coburn consented to the order and agreed to pay $300,000 in disgorgement plus $13,000 in prejudgment interest and a $75,000 penalty.

- In October 2019, the SEC filed a complaint against Telegram alleging that the company illegally participated in an ICO to raise capital from investors to finance its business.[116] The company was forced to return $1.2 billion to investors and was fined $18.5 million in civil penalties.

- In December 2020, the SEC filed a controversial lawsuit against Ripple Labs Inc. and two of its executives, alleging that they raised at least $1.3 billion from investors in the U.S. and worldwide through an *unregistered* ICO.[117] Shortly after the announcement, XRP was delisted from most major crypto exchanges.

However, Clayton's efforts to crack down on cryptocurrency can be considered mild compared to the SEC's reaction to cryptocurrency since he vacated his chair. Just when the crypto community thought things would get better with the new chairman, they didn't.

Things Got Worse

Gary Gensler, the former chair of the CFTC under the Obama administration, was sworn in as SEC chair in April of 2021. Gensler is scary, but not in the way that you might think. It's not his demeanor or the power he wields as the SEC Chair that makes him scary. In fact, he's probably one of the nicest guys in Washington. What makes him scary is that he knows cryptocurrency. Like, he *knows* it.

Gensler taught Bitcoin and blockchain at MIT's Sloan School of Management, a school whose faculty and students largely support blockchain and cryptocurrency initiatives. He is intimately familiar with blockchain, cryptocurrency, smart contracts, DeFi, and everything else associated with how the space works. That could make him one of crypto's greatest allies or its greatest threats.

His knowledge of cryptocurrency bought him significant credibility in the crypto community. When rumors of Gensler possibly becoming the next Chairman of the SEC began to circulate, the community was excited. We counted him as one of *us*, a Cryptonite, the invisible hand that would move crypto and blockchain technology into further adoption. Unfortunately, that was not the case. It was quite the opposite.

Since Gensler took office, the SEC has taken a more aggressive stance toward cryptocurrency and become more enforcement-oriented.

■ According to a 2021 survey by Cornerstone Research, the

SEC's cryptocurrency enforcement activity heightened from the end of May to mid-September 2021. Crypto companies saw nine enforcement actions in Q3 alone, including litigation and administrative proceedings, before slowing down in Q4.[118]

- In March 2022, the SEC Division of Examinations announced digital assets as one of their priorities for 2022, specifically examining the… *offer, sale, recommendation, advice, and trading of crypto-assets*, and examinations of mutual funds and ETFs offering exposure to crypto-assets.[119]

- In May 2022, the SEC added 20 new positions to their *Crypto Assets and Cyber Unit*[120]—the unit responsible for identifying crypto projects liable to enforcement action—bringing the total number of positions to 50 (40% increase). The unit is part of the SEC's *Division of Enforcement.*

Gensler has also been outspoken about crypto in his short time as SEC's chair, touting that more SEC enforcement is needed in the space. During the Aspen Security Forum in August 2021, Gensler pleaded with Congress for more authority over crypto, stating:[121]

> *"This asset class is rife with fraud, scams, and abuse in certain applications."*

> *"There are some gaps in this space, though: We need additional Congressional authorities to prevent transactions, products, and platforms from falling between regulatory cracks."*

I can't say that the community wasn't disappointed by what seemed like an all-out assault by the SEC on the crypto industry. The consensus was that, though there were clearly some bad actors in the space, anyone could be considered a *bad actor* when there are no clear, applicable guardrails for entrepreneurs to operate within.

For example, in September 2021, the SEC issued Coinbase CEO Brian Armstrong a Wells Notice (a letter from the SEC informing a person or organization on the intent to bring legal action) a week before launching their first crypto-lending platform.[122] This was significant because 1) tens of millions of traders worldwide use Coinbase, and 2) Coinbase was the largest crypto company that the SEC had gone after—so large that the outcome of the situation would probably set a precedent for the rest of the crypto industry.

Armstrong, however, didn't take it lying down. He called out the SEC in a 21-thread tweet:[123]

- *"We're being threatened with legal action before a single bit of actual guidance has been given to the industry on these products…"*

- *"We're committed to following the law. Sometimes the law is unclear. So if the SEC wants to publish guidance, we are also happy to follow that…"*

- *"If you don't want this activity, then simply publish your position, in writing, and enforce it evenly across the industry."*

I think this illustrates the level of frustration the crypto industry has experienced in wading through the regulatory ambiguity. If Coinbase, one of the oldest and most overly-KYC-AML-enforcing crypto exchanges in the financial services industry, can't avoid scrutiny from federal regulators, who can? Unfortunately for Coinbase, they scrapped the lending platform plan out of fear of legal action.[124]

Why All the Regulatory Shade?

So, why is the SEC coming after crypto companies and entrepreneurs when no clear regulations have been established? I wish I could speculate on why it's happening, but I can tell you what that general consensus in the crypto community is:

1. The SEC is trying to strongarm crypto entrepreneurs into bending to their will to set an example for the rest of the crypto industry.

2. The SEC is targeting more prominent crypto companies (think Coinbase and Ripple) because they present the most threat to the government's ability to control the flow of money in and out of the country.

Yes, the strongarm method will naturally result in the SEC—and maybe other regulatory bodies—making tons of enemies in the crypto industry, but even worse, it will result in a host of unwanted, negative, unintended consequences:

Mass confusion. Frankly, without compatible regulatory guardrails, no one knows what to do. No one will know if they are participating in an unregistered security sale, violating a regulatory standard by designing a new type of crypto product, or if the tokens they are buying and selling violate a securities standard.

Discourages innovation. No one will want to innovate out of fear of being sued by the SEC or another regulatory agency.

Drives innovation away from the country. If innovators and entrepreneurs feel threatened by federal regulators, they will simply set up shop in crypto-friendly countries like Switzerland, Germany, Singapore, Portugal, and El Salvador.

Discourages entrepreneurs from entering the country. Like the point above, entrepreneurs living outside the country will be afraid to bring their innovation into the country because no one likes to get sued by the government.

Sends the wrong message. No one likes a bully. Imposing aggressive enforcement actions against the crypto industry without first providing clear regulation makes the SEC (and possibly other federal regulators) look uninterested and unwilling to work with people who want to innovate.

But there may be hope.

In March 2022, President Biden signed an Executive Order[125] out-lining a whole-of-government approach to addressing the risks and benefits of adopting cryptocurrency and blockchain technology. Specifically, the order calls for federal agencies and regulators to ad-dress six key areas:

1. Consumer and Investor Protection

2. Financial Stability

3. Illicit Finance

4. U.S. Leadership in the Global Financial System and Economic Competitiveness

5. Financial Inclusion

6. Responsible Innovation

The crypto community was happy to hear this news because it showed that the White House took cryptocurrency seriously. However, despite the good news, it will take a significant amount of time to complete and implement. In the meantime, the industry will still be walking on eggshells, hoping not to break a rule that will place them on the wrong end of an SEC lawsuit.

So, Where Does the Answer Lie?

Probably somewhere in between.

Regulation can be beneficial to the crypto industry. Regulation can bring legitimacy to a space and capture consumers' confidence. People like regulation. People find comfort in knowing that the products and services they consume have been thoroughly reviewed by a federal regulatory body and obtained their stamp of approval.

Can you imagine what life would be like if the FDA did not regulate the medicine you take or the toothpaste you use? What if the NHTSA and the FAA did not regulate the car you drive or the airplanes you fly in? How safe would you feel? Probably not very—you'd live every day worrying that it's your last.

However, just because something *can* be regulated doesn't mean it *should* be regulated. Likewise, just because something should be regulated doesn't mean that it should be regulated to the point that it stunts innovation and excludes segments of the population that stand to benefit the most from it. And this fact is part of the issue that underpins the crypto regulation issue, leading to more questions and tension.

From a common-sense perspective, the right approach would be for regulators to work with the community to design a framework that can fit comfortably around the cryptocurrency question in a way that benefits both sides. But to say that all cryptocurrencies are securities for the sake of it demonstrates a lack of due diligence on the part of the regulators and is not fair to developers and investors.

Additionally, trying to finagle current regulatory frameworks so that they fit around the cryptocurrency question is guaranteed to make things worse. They say *you can't fit a square peg into a round hole*. I would add that if you insist on jamming the square peg into the round hole then you risk breaking one or both of them.

On a positive note, there has been a tremendous push from the crypto community to include crypto advocates in the Senate and Congress to obtain regulatory clarity. More lawmakers recognize the importance of this innovation and advocate for nationwide adoption of cryptocurrency and freedom of innovation for entrepreneurs in the crypto industry.

In 2021 alone, Congress introduced 34 new crypto-related bills (excluding the infrastructure bill). Many states are getting in on the action by introducing and passing crypto-friendly bills.

Cryptocurrency does not have to exist as a system separate from the traditional financial system or even compete with the current one. There are opportunities for crypto to supplement or improve the current financial system if federal regulators give it a chance. As naïve as this may sound, I want to believe that the billions of dollars that financial institutions and big companies invested into cryptocurrency was more than just a trendy opportunity for them to make a profit. I want to believe that there are bigger plans in the mix, especially with further innovation, but we risk not finding out unless federal regulators present sound and appropriate regulations.

It might take time, and there will be some growing pains, but the result will mean keeping innovation inside the country while at the same time ensuring investor protection. Both sides win!

CHAPTER SEVEN

BILL ON THE HILL
Control Freaks and Crypto Villains

I f you've been living in America for any time, you've probably heard (or know) that D.C. politics can sometimes be a little shady. Living so close to the Capitol, I've experienced firsthand the aftereffects of some of those quiet backroom deals that occur out of the public's eye—which are pretty common around these parts. But sometimes you see something so obviously fishy that you wonder if they were trying to mask the truth at all!

In the case of crypto, the shadiness can be found in H.R. 3684, Division H, Title VI, Section 80306—*Information reporting for brokers and digital assets.*[126]

On November 15, 2021, President Joe Biden signed into law H.R. 3684, *The Infrastructure Investment and Jobs Act*—an estimated $1.2 trillion infrastructure deal that's intended to provide Americans with good-paying jobs, rebuild roads and bridges, invest in high-speed trains, modernize hospitals, provide clean drinking water, ensure that Americans have access to high-speed internet, combat climate change, and more—all over the next decade.[127] The deal was to be part of an estimated $3 trillion social and environmental spending plan, which also included President Biden's controversial $1.7 trillion (estimate) Build Back Better Plan to improve other social and environmental issues.

Now, I'm not the liveliest person in the mornings. My emotions stay pretty baseline until I warm up to the life around me. Most mornings I crawl out of bed, don my blue pinstripe robe, get a bowl of cereal and walk around in a zombie-like state for about 30 minutes until I ease into my right mind. But I do rely on the morning news to get me going. I like to hear the weather and traffic reports and catch up on events I missed from the previous day. But on one fateful August 2021 morning I nearly chucked my Cheerios when a reporter announced that Congress had slipped a crypto tax provision into the infrastructure bill. The only thing I could think was: how bad is it?

"Broker"

The Senate drafted the infrastructure bill in response to the nation's crumbling infrastructure and the rise in unemployment resulting from the COVID-19 pandemic. It's no secret that America's infrastructure is outdated and due for an upgrade. Most of the country's current infrastructure was built in the '50s and '60s and is failing to accommodate our rising population. According to the American Society For Civil Engineers' 2021 Report Card For America's Infrastructure, the country scored a C- in its overall infrastructure, up from a D+ in 2017 and the highest score it's received in two decades.[128]

This is not surprising. I've traveled throughout the country for years for work, and I can tell you with absolute certainty that some cities (like Baltimore) have roads so bad that if you drive faster than 10 mph on them, they'll throw your car out of alignment. The Metro rail commuter stations throughout D.C., Maryland, and Virginia have concrete falling from the ceilings and occasional subway fires and train derailments. Wildfires consumed millions of acres of land across California, Texas, and New Mexico, while VA hospitals across the country are in appalling conditions. So, yes, I do agree that America's infrastructure needs a facelift. But how did a cryptocurrency tax provision end up in an infrastructure bill?

I've read several versions of how the provision found its way into the bill, but the consensus in the crypto community is that on one dark and stormy night, a cabal of regulators from the Treasury Department gathered in the bowels of the Treasury building in D.C. and drafted language placing tax reporting requirements on individuals involved in almost every facet of cryptocurrency. After the provision was drafted, it was said to have made its way into the infrastructure bill under the noses of several crypto advocates in the Senate.

The controversy over the tax provision, however, was fixed on the vague, all-inclusive definition of a cryptocurrency *broker*. The original language in the bill, crafted by the Treasury, defined *broker* in a way that could include members in the digital asset space who would not have access to the kind of trader information the IRS required them to report.

SEC. 80603. INFORMATION REPORTING FOR BROKERS AND DIGITAL ASSETS; CLARIFICATION OF DEFINITION OF BROKER

"Any person who (for consideration) is responsible for regularly providing any service effectuating transfers of digital assets on behalf of another person."

This is a pretty broad definition and can consist of non-custodial members like miners, validators, hardware and software developers, coders, or anyone else they deem fitting the definition. It was a blow to the crypto community and caused a lot of confusion and panic. How can someone report information to the IRS if they don't have access to it?

It's a known fact that the government is not friendly to people who fail to comply with the IRS. Sometimes they are thrown into jail. The threat of punitive action keeps citizens in line and on time when tax day arrives. Granted, some folks are punished because they purposely try to avoid dealing with the IRS, but this situation is fundamentally different in what the Treasury is asking folks to do: reporting information that is impossible for them to obtain, ultimately setting them up for failure.

The Dangers of the Tax Provision

From a concerned citizen standpoint, this tax provision is scary, and I fear that the Treasury and other participants of the provision lack insight into some of the long-term consequences this legislation will have on the crypto industry in America and financial innovation overall:

It Will Send the Crypto Industry to its Death. Industry entrepreneurs who want to comply with U.S. tax laws will have no choice but to discontinue operations because they will not have access to the tax information the IRS seeks. Crypto hardware equipment manufacturers will cease operations, people will lose jobs (with no new ones created), transactions will go unprocessed, new wallets and protocol innovations will go undeveloped, and the industry will collapse over time.

It Puts America at a Competitive Disadvantage. Countries that welcome crypto innovation will enjoy the benefits thereof. They will enjoy the financial benefits, the innovative use cases, the entrepreneurship, and the jobs that the technology will create. Additionally, people typically go where the money is, and if that means that crypto businesses must move their technology out of the country to operate, then they probably will.

Adds to Legislative Injustice in America. There is already a plethora of state and federal laws that unjustly target vulnerable groups of people in America. As written, the tax provision unjustly requires specific groups of individuals in the crypto industry to report information the government knows they do not have access to.

The consensus from the crypto community is that the whole thing seemed shady. We question whether this was another attempt by the powers-that-be to throw shade on cryptocurrency and collapse the industry. Or was it motivated by something else? It isn't hard to find at least one of the answers, it's about the money.

Money, Money, Money

The bipartisan Joint Committee on Taxation estimated that the crypto tax provision would bring $28 billion of tax revenue ($27.97B, to be exact) over the next decade to pay for the debt created by the infrastructure bill.[129]

Most people in the crypto community agreed that this number could not have been accurately calculated. The government doesn't even have a working regulatory framework to guide the future of the crypto industry in America so how can they accurately estimate how much money crypto businesses would bring in? Likewise, the language within the tax provision is self-defeating. The Treasury won't be able to collect $28 billion in crypto tax if the crypto industry is annihilated by same reporting requirements outlined in the provision. A number of critics believe that the $28 billion was probably one of several artificial numbers used by the Senate to justify passing the bill, nothing more.

Conspiracy theories aside, the crypto community including crypto advocates in the Senate and the House, knew something was odd about the entire tax provision and refused to let it slide. A war was afoot, and the crypto community geared up for one of the greatest battles in crypto history.

A Road Less Traveled, a Battle Harder Fought

The road leading to the final signing of the bill was not pleasant. It was three and a half months of exhausting controversies, compromises, debates, delays, and surprises as the Senate and House fought to pass one of the most extensive spending plans in American history. Within those three and a half months was probably one of the most significant ten days in cryptocurrency's history.

The crypto community's initial fight to amend the crypto tax provision waged from August 1 to August 10 and involved members from all over the community—from your average trader to pro-crypto lawmakers—going toe-to-toe with the U.S. Senate and Treasury Department to attain fair tax reporting requirements for non-custodial crypto entrepreneurs. News of the battle spread to every corner of the country, and for the first time, the topic of cryptocurrency dominated the Senate floor.

March 31
President Biden introduced the American Jobs Plan (later changed to, "The American Jobs Act").[130] which he called a *"Once-in-a-generation investment in America, unlike anything we've seen or done since we built the interstate highway system and the space race decades ago."* [131] The plan was said to be a $2 trillion plan, which would be paid for by changing the tax plans for corporations and the wealthiest Americans.

April 7
President Biden held his first conference on the American Jobs Act and highlighted the need for America to not only update its infrastructure but also to modernize it so that America could compete with other parts of the world like China.[132] His one exception, he would not impose any tax increase on citizens making less than $400,000 per year.

April 13
The Joint Committee on Taxation Chairman and Senate Finance

Committee member, Rob Portman of Ohio, announced in a hearing of the Senate Finance Committee that he was working on a bipartisan bill to close the tax gap on cryptocurrency.[133] At the time, this was something separate from the infrastructure bill. It was not exactly clear what the language in the bill would look like or when it was to be presented to the Senate for voting, but the bill was intended to impose stricter tax information reporting on cryptocurrency traders and to better define it for tax purposes.

Under normal circumstances, this would have been big news, but we (the crypto community) have heard this one before. Lawmakers have promised that crypto tax reporting laws were coming for years, yet nothing has materialized. In hindsight, this announcement was likely the origin of the idea of the crypto tax provision.

Mid-April—Late July
After President Biden's news conference on the plan, there were a lot of back-and-forth negotiations between Democrats and Republicans on how much should be spent on the plan.[134] The contention point was deciding how to pay for the plan without pinning it all on big corporations and wealthy Americans.

July 28
A spending agreement was finally made on the plan (renamed the *Bipartisan Infrastructure Deal*) and it made its way up to the Senate.[135]

August 1 – 3
The Senate held a weekend session and finally released the bill—which included the crypto tax provision—to the public. The tax provision naturally blindsided the crypto industry because no one expected crypto language to be in an infrastructure bill. Subsequent protests of the tax provision flooded Twitter and Reddit. Likewise,

crypto advocates, lawmakers, and CEOs of major crypto companies expressed their objections.

The community was urged to contact their local and state representatives to explain how the language in the tax provision would mean the end of the crypto industry—and they did just that. Congressional men and women across the country received hundreds of messages and tweets from the community. The Executive Director of the Blockchain Association in D.C., Kristin Smith, wrote an open letter to the Senate asking them to reconsider the language in the provision and their definition of *broker*, as it was detrimental to the crypto industry.[136]

The uproar from the community lit a fire under crypto advocates in the Senate, and they began to draft amendments to the tax provision.

There were about a thousand theories as to why the Treasury defined *broker* in the way they did. The top three were:

1. Policymakers at the Treasury were egregiously uneducated about how transactions on a blockchain worked.

2. It's all about collecting tax revenue.

3. The tax provision was a malicious attempt by the Treasury to discourage participation in the crypto industry because it threatened the government's control over the nation's financial system.

Whatever the reason, the community relied on pro-crypto advocates in the Senate to take the fight to the Senate floor.

August 4

Senate Finance Committee Chairman Ron Wyden of Oregon, Senate Financial Innovation Caucus co-chair Cynthia Lummis of Wyoming, and Senate Banking, Housing, and Urban Affairs ranking member Pat Toomey of Pennsylvania, put forth a compromise to the tax provision in an amended they called the *Wyden-Lummis-Toomey Amendment*.[137] The amendment provided language restricting the definition of *broker* to exclude non-financial intermediaries like miners, validators, hardware and software developers, and protocol developers. We in the community saw this as the "best-case scenario" if the Senate insisted on including language in the bill about cryptocurrency.

We were crossing our fingers and our toes that when the bill went to the Senate floor the next day, it would receive a favorable outcome. We figured that since non-financial intermediaries could not meet the reporting requirement outlined in the tax provision, the Senate would indeed exercise some common sense and give them a pass.

August 5

If you asked me to describe what happened on August 5 using an emoji, I would choose the guy with his palm to his face. On this day, Senators Rob Portman of Ohio, Mark Warner of Virginia, and Krysten Sinema of Arizona proposed an amendment of their own that would only exclude miners from the tax provision and no one else. Their amendment was meant to be a "compromise" between the tax provision and the Wyden-Lummis-Toomey amendment, but their amendment didn't seem to make sense.

For one, most crypto projects, especially the newer ones, do not use a proof-of-work consensus mechanism because its process is too slow and can be seen as environmentally taxing. Even some current proof-of-work blockchains like Ethereum are switching to faster, eco-friendlier consensus mechanisms like proof-of-stake, so excluding only proof-of-work projects will do little to mitigate the destruction the tax provision will cause.

I think the crypto community saw this curve ball coming. Senator Portman (again, who happens to chair the Joint Committee on Taxation) has been bolstering his intent to close the cryptocurrency tax gap since April, and the IRS has been promising to crack down on crypto tax evaders all year—something was bound to cast shade over the Wyden-Lummis-Toomey amendment.

Senator Lummis came out on Twitter to give her thoughts on the counter-amendment:

🐦 *"Our [Wyden-Lummis-Toomey] amendment protects miners as well as hardware and software developers. The other does not. The choice is clear."*

The community felt the same way. It was becoming more apparent that the people writing these anti-crypto laws were either very uneducated about how blockchain technology worked or simply had it out for crypto and wanted to stall its progress. It may have been a little bit of both. It was a bit of a disappointment. The last thing we wanted was another obstacle to overcome. Nevertheless, we remained hopeful and leaned on our Senate advocates to forge ahead.

August 8
The Senate voted in favor of a modified version of the bill 68–29,[138] but it did not include the modifications to the crypto tax provision

that everyone hoped for. It was disappointing, but what brought us comfort was knowing that there was still a little bit of sanity left in the Senate:

We should ensure that people aren't trying to avoid taxes by sheltering their money in digital assets, but we can do that without stifling innovation and choosing winners and losers.
                                        —*Senator Lummis, Twitter, August 8*

Though things looked bad, hope was not lost. The Senate still had about 30 hours to debate the issue before the final vote. However, the fact that the bill went forward with the original language made either crypto amendment more challenging to pass. The amendment would now need to receive a *unanimous consent vote* from all 100 Senators on the floor even to be considered for a final vote. The clock was ticking. The community went to work.

August 9
The unanimous consent vote was to take place on the afternoon of August 9. This was the last opportunity for lawmakers to agree on moving forward on the amendment and making the necessary changes to the tax provision language before it passed the Senate, so the pressure was on.

In a very odd turn of events, Senators Toomey, Lummis, Portman, Warner, and Sinema put forth a compromise to their amendments and proposed a third amendment that excluded proof-of-work miners, proof-of-stake validators, and software developers from the definition of *broker*.

Even more odd was Treasury Secretary Janet Yellen's support for the language in the new amendment. It turned out that the Senators met quietly with the Treasury to broker a compromise sometime before the vote to strike a deal (I told you at the beginning of this chapter that these backroom deals happen a lot!).

There was no doubt that the compromise resulted from the crypto community banding together to mass-protest the tax provision, and the people who supported it. What the community achieved in only a few days was astounding. The compromise wasn't all we hoped for—the vague definition of *broker* could still be open to misinterpretation—but it was 100% better than the original language. It gave the crypto industry more room to work with.

With all the bipartisan support from the Senate and buy-in from the Treasury, there was no way that this amendment would fail...at least that was what everyone thought.

Before the Vote. Before the unanimous consent vote, Senators were allowed to plead their case for the amendment on the Senate floor. Of course, Senators Lummis and Toomey took turns at the lectern trying to talk some sense into the congregation of Senators, but then, something that I didn't expect happened. Senator Ted Cruz, one of the most controversial legislators in the bunch, charged to the Senate floor to plead his case for cryptocurrency, noting that the tax provision's definition of *broker* would "obliterate cryptocurrency."

I can't say I'd seen him speak at length before, but he's a pretty dramatic dude. Comically dramatic. He yells, makes condescending (yet funny) jokes and says some weird things. But when it came to cryptocurrency, he understood its potential for the country:

"There is a new and exciting industry in the United States of cryptocurrency, whether Bitcoin or otherwise, that is generating jobs, entrepreneurs who are creating new values, new hedges against inflation, new opportunities, and it is fast moving. It is dynamic."

I was initially confused as to why he was on the floor defending crypto anyway—he wasn't one of the signatories on any of the amendments—but then it occurred to me that his state of Texas is one of the country's largest hubs for cryptocurrency mining companies—including several which relocated from China during the great crypto purge of 2021.[139] (We will discuss this at length in the chapter on China.) He probably had more reason than anyone to quash the crypto tax provision because there were a considerable number of jobs in his state on the line.

As expected, he pleaded with the Senate to consider one of the two crypto amendments Lummis et al. put forth, but he also proposed his own amendment—an amendment to strike the crypto tax provision entirely from the bill.

"My amendment is very simple. It doesn't add anything new to this bill. It just strikes these provisions, [it] says, "Look, let's not do this until we know what we're talking about. Let's be cautious. Let's be reasonable. Let's not be the number one economic developer for the Communist Party of China by sending cryptocurrencies overseas to our competitors because we've made it impossible for them to succeed here."

It was a bold move considering some Senators staked their reputations on the $28 billion estimated tax revenue that IRS allegedly would collect to pay for the infrastructure plan.

The community hoped that Senator Cruz and the other speakers made a convincing enough argument to gain unanimous consent to pass the amendment. We were optimistic about a favorable outcome but were keenly aware that the vote could go sideways.

The Vote. That afternoon, the Senate voted on the amendment. The vote result came shortly after; it was 99-1, and the crypto amendment failed to pass the Senate. The reason the amendment failed, one man, Alabama Senator Richard Shelby. Senator Shelby employed a tactic to derail the amendment by attaching his own amendment, increasing defense spending, to the vote. It came out of nowhere. His amendment was subsequently blocked by Senator Bernie Sanders, who is known to oppose increases in defense spending. And because Senator Shelby's defense amendment was rejected, the crypto amendments were also rejected.

Game over.

I don't think anyone could wrap their heads around what had just happened. It was so sudden, so unexpected.

The first question on everyone's mind was, *Who the F* is Richard Shelby??* The second as, *How was an underhanded tactic like that possible?* It was almost like watching an episode of Scooby Doo when Fred would snatch the mask off the bad guy to reveal his true identity. Who knew that during all this time the real crypto villain was not Senator Portman or Warner or Sinema or even the Treasury, but someone quiet, lurking in the shadows, someone whose name never came up in any of the tax provision negotiations—a ghost.

News of the decision spread quickly through crypto news sites and social media and a period of mourning commenced.

August 10

The next day, on August 10, the Senate voted 69–30 in favor of passing the infrastructure bill with its original language for the crypto tax provision intact.[140] The bill was headed to Congress for a final vote before being signed into law by President Biden.

A nine-day demonstration against legislative injustice ended with a seeming defeat; a defeat based on something of a technicality. Not a fun way to go out. We were down, but we still had some fight left in us. At that point, our best option was to intercept the bill in the House.

The New Crypto Winter – No Word from the House

The infrastructure bill moved through the House at a snail's pace compared to how it moved through the Senate. For months, there were voting delays and back-and-forth debates between both sides of the aisle. The House focus was almost exclusively on ratifying President Biden's Build Back Better Plan and closing the tax loophole on crypto "wash sales."

Meanwhile, members of the crypto-community were trying to lift each other's spirits as they weathered months of silence from the House on the crypto tax provision. It was like a new kind of crypto winter. You could tell that the energy wasn't the same as before. Watching all the hard work that everyone had put into amending the tax provision come tumbling down in the Senate was emotionally taxing for the community.

Nevertheless, the community appealed to advocates in the House to take another stab at amending the crypto tax provision. It was the last real chance of getting it done before the president signed it into law.

The requests did not fall on deaf ears. The House's bipartisan Blockchain Caucus Co-Chair, Tom Emmer, a beloved crypto-community advocate, sent a letter to every member of the House of Representatives on the crypto-community's behalf expressing his concerns about the language in the bill, underscoring that the language needed to be amended. In addition, crypto advocates and blockchain organizations once again lobbied the House to take action on the tax provision. But all attempts failed.

It was clear there was little to no interest in amending the provision, and on November 5, the House voted to pass the Infrastructure Investment and Jobs bill as is. Ten days later, President Biden signed the bill into law.

We Got Beat, But Did We Lose?

It would not be fair of me to say that no good came out of this experience. Of course the community was disappointed with the outcome. However, looking at it from a different perspective, this experience did more good for the community than people may realize:

Unity in the Community. The initial rally by the crypto community in August to protest the bill's first draft was a testament to their commitment to the crypto industry and each other. Individuals in the community banded together to fight for their brothers and sisters in the community who would be negatively affected by the tax provision.

Utilizing the Democratic System. The crypto tax provision compelled individuals in the crypto community to open a dialogue with their local and state representatives. For many, it was their first time contacting their Congressman/Congresswoman and Senator about issues that affected them or someone else. This behavior is good for democracy, and they hopefully find that working with their representatives was a worthwhile mechanism to effect change.

Education for Policymakers. As the community pushed their representatives to rectify the misleading language in the crypto tax provision, they were simultaneously educating them on matters of blockchain technology and digital assets that they probably would not have received otherwise.

National Recognition. Probably the most valuable outcome of this experience was the recognition cryptocurrency received from mainstream audiences. When the crypto community unified to protest the tax provision, it made national news and shed light on a piece of legislation that unjustly targeted a vulnerable group of people. This ultimately attracted more supporters to the cause and attested to the crypto community's belief in the technology.

The Fight Continues

Okay, by now you probably think this tax provision is the beginning of the end of cryptocurrency as we know it, but this may not necessarily be true. I don't want to end this chapter on a bad note, so I will give you one more message of good news—the tax provision will not go into effect until 2024, which gives the community more time to lobby for changes. In the meantime, the community is focused on getting clear guidance from the Treasury on who will be considered a *broker* when tax time comes.

In December, a letter[141] endorsed by six Senators, including Lummis, Portman, Toomey, and Warner, was sent to Janet Yellen urging her to provide information or informal guidance on the definition of *broker*. This is critical because some legal language can be interpreted a hundred different ways, depending on who's doing the interpreting. It's better to err on the side of caution and get something concrete into law to prevent future administrations from misinterpreting the provision and putting forth regulations that will further damage the crypto industry.

Time will tell how all of this will shake out, but until then, we continue to work with our leaders in the Senate, House, and the crypto community and share the benefits crypto brings to the world.

CHAPTER EIGHT

IF YOU CAN'T BEAT 'EM, BAN 'EM
The Federal Inferiority Complex

The need for control can have catastrophic effects when wielded by someone who feels like they cannot manage life without it. Needing to control people or situations can come with a heavy burden and, when exercised inappropriately, can cause unnecessary suffering for oneself and others.

Oftentimes, the need to control is rooted in some fear, like the fear of losing status or losing a prestigious position at work. It can be the fear of disappointing others, being looked down upon by peers, or disappointing oneself. People can also fear abandonment and fear getting hurt.

Next to love and greed, fear is probably the greatest motivator there is, and some people will manipulate the actions of people and situations to prevent a feared event from materializing.

Some people have strong views on how things should be, and they spend time, money, and resources making sure that their version of the status quo is achieved and maintained. The problem is that when these people are in a position where they make decisions for others, they often have knee-jerk reactions to changes that threaten their views of how things should be and end up making things worse for everyone. Fear is a powerful force.

Unfortunately, not even decision-makers in the world's governments are impervious to fear's power to influence their decision-making. This is especially true when it comes to global cryptocurrency adoption.

Cryptocurrency's greatest advantage is its ability to give people sovereign control over their money and financial futures. As I explained in my chapter on the Ethereum Experience, crypto projects offer people various financial services that are separate from the government's financial system.

With these services, there are no high barriers of entry into the crypto market as there are with traditional equity markets. There are no lengthy loan applications or credit checks, you don't have to fork over a lot of unnecessary personal private data to the lender, and there are no discriminatory financial practices based on race, class, or gender—something that we see in almost every financial institution in the world.

And because crypto projects run on a blockchain—where the public can see and verify every financial transaction made—it substantially reduces a person's (or institution's) abilities to bury fraudulent transactions where no one can find them. This is largely why some nations ban crypto.

Implicit Vs. Explicit

But before we move ahead, let's look a little bit closer at what "banning" is in the context of a cryptocurrency ban. I like how the Law Library at the Library of Congress defined the two different styles of bans (explicit and implicit) because it captures how countries approach the cryptocurrency question.[142]

....

An explicit ban means that a country has prohibited citizens, financial institutions, companies, and exchanges from having any involvement whatsoever with cryptocurrency. This includes buying, selling, borrowing, lending, holding, you name it. Nepal, for example, has declared cryptocurrency and crypto mining illegal through their 2019 Foreign Exchange Act. Any citizen found to have bought, sold, traded, mined, or even encouraged cryptocurrency will be punished per "existing laws."

Conversely, implicit bans will allow citizens to trade and hold cryptocurrency but prohibit financial institutions and exchanges from participating in any cryptocurrency-related activity or offering cryptocurrency-related services. For example, Southeast Asia's largest economy, Indonesia, prohibited financial service institutions from trading, marketing, facilitating trading, and using cryptocurrency as a form of payment. However, the country will still allow cryptocurrency to be traded as a commodity on a futures exchange.

....

The Big Bans

The same Law Library report found that 51 countries worldwide have placed bans on cryptocurrency.[143] This came as a shock because I didn't think 51 countries cared enough about cryptocurrency to even ban it. But I suppose it does make sense that almost two-thirds of those bans occurred between 2020 and the end of 2021, given Bitcoin's explosion in the latter part of 2020 to 2021.

There's no question that the rise in crypto's popularity during this time frame gave central banks a cause for concern. Nine of those 51 countries have placed all-out explicit bans on cryptocurrency:[144]

Algeria	Egypt	Nepal
Bangladesh	Iraq	Qatar
China	Morocco	Tunisia

The other 42 have placed implicit bans on it:[145]

Bahrain	Chad	Georgia
Benin	Congo	Guyana
Bolivia	Côte D'Ivoire	Indonesia
Burkina Faso	Democratic Republic of Congo	Jordan
Burundi		Kazakhstan
Cameroon	Ecuador	Kuwait
Central African Republic*	Egypt	Lebanon
	Gabon	Lesotho

Libya	Nigeria	Turkey
Macao	Oman	Turkmenistan
Maldives	Saudi Arabia	United Arab Emirates
Mali	Senegal	Vietnam
Moldova	Tajikistan	Zimbabwe
Namibia	Tanzania	
Niger	Togo	

Ban has been lifted since the report.

Let's take a look at some of the biggest crypto bans.

China

China was the first notable country to ban cryptocurrency. China's ban was an exceptional case because 1) They have the second largest retail market in the world, and 2) In addition to banning cryptocurrency, they banned cryptocurrency mining, though before the ban, approximately 65% of the world's Bitcoin mining took place in China.[146]

China's cryptocurrency restrictions date back to 2013, when the Central Bank of China prohibited financial institutions from trading bitcoin.[147] Their recent bans of cryptocurrency and cryptocurrency-related activities like mining had a cataclysmic effect on Bitcoin's price and the crypto industry. I've dedicated the next chapter to China's cryptocurrency banning because that's just how epic it was.

Turkey

Turkey released their *Regulation on Non-Usage of Crypto Assets in Payments* in April 2021[148] as the country has been struggling with a si-

lent financial crisis. Shortly before the ban, Turkish president Tayyip
Erdogan fired the head of his Central Bank, Naci Agbal, causing the
value of the Turkish lira to decline by 15%.[149] Agbal was the third
Central Bank of Turkey head to be fired in four years, which did
little to help Turkey's struggling economy.

Agbal's unexpected termination coupled with the lira's subsequent
drop in value led Turkish citizens to seek cryptocurrency to hedge
against Turkey's inflation. The Turkish government has been aware
of Turkey's growing cryptocurrency market since the beginning of
the crypto boom in late 2020, but it wasn't until after cryptocurrency went supersonic in 2021 that the Turkish government initiated
the ban, citing cryptocurrency's risks, volatilities, and illegal uses by
nefarious individuals.

Indonesia

In 2017, Indonesia's financial service authority (Otoritas Jasa Keuangan (OJK)) declared bitcoin an illegal form of payment,[150] citing
many of the cookie-cutter explanations other nations used: *people
will lose their money, the market is too volatile, only bad people use
it*, etcetera, etcetera. Ironically, one year after the ban, the country
allowed crypto to be traded on the commodities exchange, probably
because the Commodity Futures Trade Regulatory Agency (Bappeb-
ti) could collect tax from crypto transactions.

Since the ban, however, Indonesia has become one of the leading
countries in Southeast Asia to use cryptocurrency.[151] Crypto's popularity in Indonesia was so prominent that the OKJ reminded financial institutions via Instagram that marketing and facilitating crypto
trading was strictly prohibited.[152]

"We Don't Need Cryptocurrency"

Some people favor crypto bans because they believe the current financial system is efficient and serves everyone equally. They say, *"Why do we need cryptocurrency? We already have money and a financial system that works."*

Unfortunately, this is far from accurate. Several countries, including the U.S., have financial systems that don't work for everybody, which is evidenced by a worsening economic climate and the fact that nations have been dealing with the same economic issues for decades:

- **Global financial inequality** is still the leading disadvantage of the financial system. Most of the world's wealth is held by a few. The wealthy have access to financial resources and education, financial experts, and investment opportunities that allow them to accumulate more wealth. Meanwhile, the poor (and a portion of the middle class) who struggle to make ends meet cannot afford to invest or gain access to the same financial resources.

- Since abolishing the gold standard, **central banks have no limit** to the amount of currency they can create. There is very little oversight. They can print as much as they want, when they want. Uncontrolled printing can lead to inflation if not managed correctly.

- Uncontrolled **inflation** leads to the devaluation of savings and pensions, stock market crashes, spikes in food, gas, and electricity prices, recessions, hyperinflation, political unrest, and government coups.

- **High transaction costs** limit the flow of remittance to foreign countries, which puts financial strain on both the sender and receiver.

- **Fractional reserve** banking increases the likelihood of bank runs and insolvency, which could lead to an economic depression.

- **Greed and Corruption** – This one is not hard to figure out. Think, *2008 financial crisis.*

Governments are fully aware of deficiencies in the global financial system, and the people who claim there is nothing wrong with it are the ones benefiting the most from it. All the while, low or middle-income families who live in countries that impose heavy taxation and corrupt monetary policies struggle to get by.

Cryptocurrency has become so prevalent in these affected countries and communities because it frees people from the burden of operating under the global financial system. That kind of freedom loosens the government's control over them.

Case Study – Nigeria

Nigeria is probably one of the most underrated countries in terms of technological competencies. A large portion of Nigeria's population are young, intelligent, tech-hungry men and women who are ahead of the game in understanding the transformational value of technologies like cryptocurrency.

Nigerians have fully embraced cryptocurrency in recent years and see it as a tool for overcoming economic and monetary barriers that have previously prevented them from building a quality life or business. They've used it to start and expand businesses, generate wealth through investing, and send funds to friends and family members in need at a cheaper and faster rate. They also use it to protect their savings from the country's failing currency, the *naira*, and to combat government corruption.

For example, in March 2019, when the Central Bank of Nigeria placed textiles and textile materials on a list of items that were ineligible for foreign exchange, Nigerian business owners looking to import goods from overseas trading partners could not purchase the U.S. dollars, renminbi, or euros necessary to pay their suppliers. As a result, some business owners turned to bitcoin to pay the overseas partners that accepted it[153] (and many of them did!).

Likewise, during the 2020 *EndSARS* protest to end police brutality in Nigeria, Nigerian relief organizations began accepting donations in bitcoin to pay for water, food, and medical supplies for protestors after the Nigerian government suspended their bank accounts and fiat donation websites.

Crypto adaption became so prominent in Nigeria that the country ranked as one of the world's leading countries for bitcoin trading in 2020.[154]

The liberation that cryptocurrency brought to Nigerians, however, began to upset the natural order of things, specifically the revenue the government collects from remittance payments.

In Nigeria, remittance inflows, that is, money sent back to families in Nigeria from friends and family members living abroad, made up about 6% of the county's GDP, according to 2018 data from

PWC.[155] In 2020, remittance inflows to Nigeria dropped 27% (down $6.3 billion from the previous year) despite an overall 2.3% increase in remittance inflows to Nigeria's neighboring countries in Sub-Saharan Africa.[156]

The COVID-19 pandemic, which curtailed most of the world's economies, was partly to blame for the slowdown in remittance. Still, some experts believe that cryptocurrency use by Nigerians outside the country also accounts for the steep drop. That is because Nigerians living abroad opted to send cryptocurrency back to Nigeria instead of using the government-approved fiat payment system, which Nigerians opine is notoriously slow and unreasonably priced compared to cryptocurrency.

The Nigerian government recognized the threat cryptocurrency posed to their ability to control the flow of money within the country, therefore, to put the lid back on the crypto genie lamp, the government intervened.

In February of 2021, the Central Bank of Nigeria issued a letter mandating banks and other regulated financial institutions in Nigeria to identify and close the accounts of customers and crypto exchanges that transacted cryptocurrency within their systems.[157] Before the ban, Nigerians could purchase cryptocurrency using the Nigerian naira, but after the ban prohibited Nigerians from using the naira to buy cryptocurrency, they sought other countries' currencies to purchase cryptocurrency.

Nigerians, and citizens of other like-nations, recognize cryptocurrency's positive impact on their financial situation and they are not giving up the fight. There has been an enormous push for economic reform around the legality of cryptocurrency at the federal and state level. Some governments, like the government of El Salvador, have fully embraced cryptocurrency and made it a staple in their economic systems. Others, like the United States and Canada, are still working through the details, but the one thing I think most governments agree on is that cryptocurrency is here to stay.

Crypto Competition – The Central Bank Digital Currency

Governments quickly realized that banning cryptocurrency wasn't enough to curb the public's appetite for it, so they pivoted their strategy and introduced a concept they could use in their fight against cryptocurrency, the *Central Bank Digital Currency.*

A Central Bank Digital Currency, or *CBDC,* is a virtual form of a digital fiat currency that central banks issue directly to the customer.[158] Central banks are putting on massive promotion campaigns for the CBDC, highlighting its ability to solve many problems affecting global financial systems. It's in all the magazines, on all the news programs, on the radio, and is one of the hottest subjects in the crypto industry. CBDCs are currently in the research, pilot, or use phase in most countries.

Some people assume that because CBDCs are digital currencies, they possess the same qualities as cryptocurrencies like bitcoin and ether, but this is far from the case.

First, most CBDCs will not run on an open network like Bitcoin and Ethereum, but on pre-existing payment gateways where data is stored on centralized servers and controlled by a central authority.

Second, CBDCs do not have a fixed supply, so central banks theoretically can mint as many coins as they please. As such, there is a chance that CBDCs will have the same inflationary risks as fiat currency.

Third (for now), CBDCs do not have cross-chain operability or a governance mechanism like most other cryptocurrencies. The public will have no say in how CBDCs are used, disseminated, tracked, or managed. All power will remain in the hands of the government.

CBDCs' functionalities will vary by nation, but the overall concept is the same: CBDCs will be issued directly from a central bank to the user via a downloadable digital wallet. Some central banks have even discussed CBDC bank cards as another option to store and spend CBDCs.

To date, there are almost 80 countries that are either researching, experimenting or have launched CBDCs, according to CBDC Tracker.[159]

Austria	Eastern Caribbean	Hungary
Bahamas	Economic and	Iceland
Bahrain	Currency Union (OECS/ECU)	India
Bhutan	Egypt	Indonesia
Brazil	Eswatini	Iran
Canada	Euro Area	Iraq
Chile	France	Israel
China	Ghana	Jamaica
Curaçao	Guatemala	Japan
Czech Republic	Honduras	Jordan
Denmark	Hong Kong	Kazakhstan

Kenya	Oman	Taiwan
Kuwait	Pakistan	Tanzania
Laos	Palestine	Thailand
Lebanon	Peru	Trinidad and Tobago
Macau	Poland	Tunisia
Madagascar	Qatar	Turkey
Malaysia	Republic of Palau	Uganda
Mauritius	Russian Federation	Ukraine
Mexico	Rwanda	United Arab Emirates
Morocco	Singapore	United Kingdom
Namibia	Saudi Arabia	United States of America
National Bank of Georgia	South Africa	
	South Korea	Uruguay
Nepal	Sudan	Vietnam
New Zealand	Sweden	Yemen
Nigeria	Switzerland	Zambia
Norway		Zimbabwe

CBDC Tracker data as of June 2022

Some countries are further ahead than others and are worth highlighting:

Nigeria. Nigeria became the first country in Africa to officially launch its own digital currency. In October 2021, the Central Bank of Nigeria launched the *eNaira*, a digital version of the naira. The eNaira caught citizens by surprise as there was very little buzz about it before its launch. One of the central bank's selling points for the eNaira is its cheap, quick, and easy processing of remittance payments, which is attractive to Nigerians living abroad and crucial to the government's revenue.

The United States. In January 2022, the Federal Reserve released its 40-page discussion paper[160] that examined the pros and cons of CBDCs in the United States. The Fed has not yet taken an official stance on CBDCs, but they are soliciting feedback from the public on CBDC topics such as cross-border payments, potential effects on the U.S. financial system, privacy, legal tender, and more.

China. In late 2019, the People's Bank of China (PBOC)'s Digital Currency Research Institute launched a pilot program for the digital yuan called the *e-CNY*[161] *(e-CNY – formerly known as the "digital currency/electronic payments" (DCEP) project).* China has been researching a digital version of the yuan since 2016 in hopes of being the first country to fully implement a digital currency. Once the e-CNY goes live, users can access many e-CNY-related payment devices, including smartphone-free hardware wallets, contactless cards, and wearables.

According to a 2021 PBOC e-CNY progress report,[162] e-CNY transaction volume totaled 70.75 million, approximating 34.5 billion yuan ($5.34 billion) at the end of June.

South Korea. In December 2021, The Bank of Korea (BOK) announced that they'd completed the first of two phases of mock simulation testing for their CBDC, the digital won.[163] The first phase tested the digital won's performance in a simulated cloud environment while the second phase will test its wiring and payment in an offline environment and cross-border wiring and trading.[164] The BOK hoped to complete phase two of the testing by June 2022.

Again, these are only a few examples of how some nations are progressing with CBDCs, but I see a bigger trend here: Governments are attempting to reduce the public's need for cryptocurrency by introducing their own imitation cryptocurrency.

The Scary Side Of CBDCs

In all fairness, CBDCs, if prescribed appropriately, can have valuable advantages like ease of access, ease of use, and greater financial inclusivity. However, looking at the scarier side of the CBDC, it can present several dangerous ramifications that will jeopardize privacy and economic stability.

Privacy. Privacy invasion is consumers' primary concern with CBDCs. Unlike cryptocurrencies, CBDCs do not facilitate anonymous transactions. Whereas people use paper cash to make private transactions, with CBDCs the government can track and scrutinize even the most minor transactions. Because central banks mint, disseminate, and control CBDCs, they can theoretically:

- Freeze accounts for any reason

- Remove CBDCs from users' wallets

- Block users from purchasing certain items or making certain transactions

- Identify users' purchasing patterns and create *spending behavior* profiles

- Place limitations on the amount of CBDCs one can hold

- Place a "spend by [expiration date]" on users' accounts. (Users must spend x-amount of CBDC before a specific date before the Federal Reserve removes it from their accounts. This can be especially pronounced during times of recession when the government needs people to spend to restimulate the economy.)

There are also certain macroeconomic and safety ramifications for CBDCs:

Bank Runs and Weird Competition. People may prefer depositing their funds into CBDCs rather than commercial bank accounts. Not only might this result in a progressive bank run by drying up bank liquidity, but it will force commercial banks to compete with the CBDC by offering customers additional deposit incentives. In a sense, this makes the banking system self-defeating: While central banks intend for CBDCs to improve the financial system, using them may cause other (bigger) issues in the financial system and the economy.

Centralized and Exploitable. CBDC transactions will likely occur on an existing payment system that is susceptible to hacks, viruses, outages, internal sabotage, and physical attacks. Compromises to CBDC databases will have catastrophic effects on the affected nation's economy and the economies of partnered countries.

Infrastructure and Upgrades. Lesser developed nations may not have the appropriate infrastructure to maintain and protect CBDC databases from events like physical/cyber attacks, unfavorable climate and weather conditions, and natural disasters. Similarly, they may not possess, or be able to outsource, the technical expertise required to provide significant upgrades to the CBDC database and wallet software or to maintain the network.

You Cannot Beat by Banning

The question for governments is now, *"How do you enforce a ban on a decentralized finance system?"* It's a question I believe many governments have not wholly considered, so until they can come up with a workable solution, they will most likely continue to impose hasty umbrella restrictions on cryptocurrency to slow down its adoption.

However, this approach is doomed to fail.

Governments can ban cryptocurrency as much as they please, but there's very little they can do to enforce it. Decentralized networks are built to be censor-proof—people do not need the government's permission to buy, sell, send, or hold cryptocurrency. Granted, there are vectors that governments can exploit to restrict people's access to cryptocurrency—like ordering banks to freeze the accounts of crypto traders—but as a network, they cannot shut it down. If people want crypto, they will find a way to get crypto.

This concept is difficult for some people to grasp because, as people who lived our entire lives under the rule of centralized entities, we had no choice but to use the government's tool. Until now, centralized financial institutions, which operate based on the monetary policies set by the government, made up 100% of our financial system. But there is a new system in play—an alternative to the old ways of doing things—a system in which no government can dictate policy.

I would also apply a warning about CBDCs. If you abruptly throw CBDCs—a new form of government currency that no one in the government completely understands—into the world as a solution to the cryptocurrency *problem*, consequences could be dire. And with so many disadvantages to the CBDC, if not designed and implemented correctly, it will do more harm than good.

The best approach would be for governments to work with leaders in the crypto industry to design a framework whereby both systems coexist and complement each other. Why not keep the talent and technology inside the country? It brings all the things a good government would want: jobs, capital, innovative people, and a bright future. It might take time, but in the end, it will pay off.

CHAPTER NINE

CHINA
The Gold Standard for Crypto Controversy

I f throwing shade on cryptocurrency was an Olympic sport, the Chinese government would win Gold every four years.

The Chinese government does not respond kindly to existential entities, objects, or influences entering the country and putting crazy ideas like financial independence into the minds of its citizens. And as you will soon find out, the government for decades has been obsessed with achieving economic growth and self-sufficiency to compete on a global scale and reestablish China as the economic powerhouse it once was. And because of this obsession, the government has placed cryptocurrency trading and crypto-related activities like mining in their crosshairs.

It's a Fact. The Chinese Government Hates Cryptocurrency

The Chinese government's animosity toward cryptocurrency is well known throughout the crypto community. It's become something of a thing. When you randomly ask crypto traders which country hates crypto the most, they will almost immediately respond with, "*Oh, that's easy, China.*" That's because those of us who've put a few years into cryptocurrency remember all the years the Chinese government villainized the crypto industry.

If you recall from the previous chapter, China was one of the first countries to "ban" cryptocurrency (bitcoin). In fact, the People's Bank of China (PBOC), China's central bank, takes first prize when it comes to the number of press releases targeting cryptocurrency. Let's look at their track record Since Bitcoin:

January 2009: Bitcoin goes live.

May 2010: Software designer Laszlo Hanyecz made the first purchase using bitcoin.

Bitcoin gains popularity – More people start mining and trading it.

December 2013: PBOC prohibited financial institutions from trading bitcoin.[165]

September 2017: PBOC banned Initial Coin Offerings (ICO).[166]

February 2018: PBOC announced they'd block access to all domestic and foreign cryptocurrency exchanges and ICO websites.[167]

April 2019: China's National Development and Reform Commission (NDRC) listed crypto mining in their Industrial Structural

Adjustment Guidance Catalogue as an "undesirable" industry and therefore should be eliminated from consideration.[168] *(This was later removed.)*

May 2021: China's Financial Stability and Development Committee of the State Council announced a crackdown on bitcoin mining and trading activities in China.[169]

September 2021: NDRC issued a Notice on Regulating Virtual Currency "Mining" Activities restricting cryptocurrency mining activities in the country.[170]

September 2021: PBOC issued a circular on "Further Preventing and Disposing of Speculative Risks in Virtual Currency Trading," reiterating cryptocurrency's illegalities in the country.[171]

November 2021: NDRC announced they would raise electricity prices on miners who continue to participate in crypto mining.[172]

To understand why China has taken an offensive posture toward cryptocurrency, you must understand China's economic history. Now, if you're already familiar with China's economy in the nineteen and twentieth century then feel free to skip ahead a few sections. However, if you don't mind bearing with me for the following few pages, I think taking a quick peek at China's struggles in the past will give you some valuable insight.

The Chinese People Suffered Horribly in the Nineteenth Century

China was once one of the wealthiest nations in the world during its premodern era between 200 B.C. to the early 1800s. They were strong in agricultural production. It was their thing. Before the twentieth century, China's agrarian acumen made the country self-sufficient. They had endless acres of farmable land, a robust labor workforce, and a pride-driven work ethic. Unlike other nations, China wasn't concerned with securing international trade to bolster its economy because it already had everything it needed, including plenty of land, lakes, and rivers.

At the height of its glory in the nineteenth century (the 1820s to be exact), China's economy was six times that of Great Britain, and its GDP was almost 20 times that of the U.S.[173]

These numbers are pretty impressive given the aggressive expansion into retail markets by the U.S., Germany, and Great Britain.

China's economic downfall, however, escalated during the first half of the twentieth century—the country was governmentally fragmented, communism was on the rise, the country suffered invasions and endured several rounds of civil war, and their agriculture production drastically declined, causing countrywide food shortages.

Furthermore, China's once-thriving economy took a backseat to Western countries like the U.S., Germany, France, and Britain which became industrial giants during the first and second industrial revolutions. While the U.S., Western Europe, and Japan used power-driven machinery and made advances in iron, steel, chemical, electrical, and automobile production, China still used coal-fired machines and lacked industrial innovation and productivity. They could not keep pace with modernized Western economies, and as a result, their economy suffered. The Chinese became desperately poor. Their country was physically in shambles, and families could barely feed themselves after their countryside farmlands were destroyed.

China needed help, and then came Mao.

Communism at Its Worst – Enter Mao Zedong and "Maoism"

China took a turn for the worse in 1949 when Chinese Communist Party (CCP) Chairman Mao Zedong—fashioning himself as China's best chance for a better tomorrow—declared China the *People's Republic of China* and embarked on a quest to clean up the destruction left behind from decades of wars. Mao wanted to transform China into a global superpower in order to compete with the already industrialized United States and Europe, and he wanted it to happen fast.

It was during this time that Chinese civil liberties began to deteriorate. In 1953, after the initial clean-up phase was complete, Mao put his governmental personnel into place and enlisted the help of the Soviet Union for economic assistance. He wanted to impose an economic model like that of the Soviet Union to catapult industrialization in China. It was part of Mao's five-year plan to make China a competitor on the economic world stage.

However, there was a catch. By imposing the Soviet model, the state seized all of China's private businesses, properties, and economic resources. Landlords were forced to hand over property rights, and any private ownership of capital was prohibited. Those who the government deemed as "counter-revolutionaries"—former capitalists, landlords, religious leaders, or anyone who distinguished themselves as intellectually or financially superior to the government—whether unwittingly or not—were arrested, jailed, or even killed. People were afraid to be labeled as "class enemies" and targeted by the government so they gave in willingly. And by 1956 the government owned

100% of private businesses and property in China.

Likewise, the government maintained direct control of all provinces in China, despite the existence of provincial governors. The government ultimately seized complete control over the country's capital, resources, and people.

The country saw some signs of modernization, but relations between the Chinese and Soviets began deteriorating by 1958. Mao believed the Soviet Union did not stay true to the goals of communism, so that year, China abolished the Soviet economic model. They imposed a new model whereby they gave provincial governors control over their own provinces to manage their economies as they saw fit. This was the first time that local leaders had autonomy over their territories during this era. They hoped that this new economic model would be the catalyst to propel China's economy ahead of the West.

It was a bold move, but it also meant that local leaders were held personally responsible if their province was not producing to the degree Mao intended. This put insurmountable pressure on local leaders to perform.

The plan was for farmers in the countryside to grow crops to feed industrial workers in the cities. This was so that industrial workers could work longer and produce more. But Mao's plan eventually hit a snag: China's agricultural productivity greatly outproduced its industrial productivity, which was not what Mao planned initially. Remember, Mao was extraordinarily impatient and wanted China's economy to modernize practically overnight. So in 1959, he changed course and did something that would become one of the worst human tragedies in history, a new five-year economic plan that the CCP dubbed the *Great Leap Forward*.[174]

The New Five-Year Plan

If you are not familiar with the history of the Great Leap Forward, then I encourage you to take some time to read about it. It puts into perspective the mindset of Communist China in the 1950s and 1960s and how elements of that mindset carried over to the current Chinese government.

But to summarize, the Great Leap Forward was Mao's new five-year economic plan to increase China's industrial productivity and to transform China into a communist paradise. They would not accomplish this with power-driven machinery and modernized farming equipment (like a normal society would) but through a robust workforce, manual labor, and sheer willpower. However, Mao's unrealistic production goals eventually led to massive deforestation, countrywide famine, and the deaths of tens of millions of Chinese.

Mao divided the labor force into two components: peasant farmers who oversaw tilling the farmland in the countryside and industrial workers who were in charge of producing the steel in the urban cities. Peasant farmers were tasked with growing crops to feed the industrial workers in the cities. Industrial production—steel and otherwise—was Mao's trump card. He believed that if China could ramp up its steel production, then everything else would fall into place. He also believed that industrial productivity was directly proportional to the number of crops farmers produced to feed industrial workers, therefore, he imposed unrealistic production quotas and pressured peasant farmers to produce.

But the industrial work in the cities wasn't enough for Mao. He wanted industrial production to happen yesterday. His obsession with steel led the government to supply farming communes with backyard steel burners for peasants to smelt (meltdown and purify) scrap metal. They smelted their pots, pans, spoons, forks, knives, doorknobs, shovels, practically any metal available.

This created another problem. The abundance of furnaces increased the demand for wood for fuel, which led to deforestation in the countryside. When the wood from deforestation ceased, peasants burned their furniture and doors. Peasant farmers had no training in smelting, so the steel they produced was low quality and unusable. The process was utterly unproductive, so the backyard burner plan was scrapped.

What followed was terrible weather, droughts, locust outbreaks, and bad land tilling. This caused agricultural production to suffer severely, and farmers could no longer maintain quota. But instead of reporting problems with crop production, provincial leaders who feared the scorn of the CCP reported inflated numbers, sometimes up to two- or three-times the usual quota. So, when the state arrived to collect their tax on the grain, they collected more than the communes could afford to lose, leaving workers without enough food to go around. In some cases, communes fell short of their tax obligations and were accused of hiding grain, at which point they were beaten or tortured.[175]

As the food supply dwindled, a countrywide famine struck. People became so hungry that they ate grass, tree bark, soil, and boiled leather. When that was no longer an option, they resorted to murder and cannibalism. People were dying by the hundreds. Corpses of men, women, and children were left rotting in the fields or on the side of the road.

Mao was aware of the problem but refused to seek assistance from the international community. He wanted other nations to believe he had a good handle on the country. But in reality, the country was on the verge of another economic disaster.

The government ended the Great Leap Forward in 1960, two years before planned. Historians estimated that between 20 and 45 million Chinese died between 1958 and 1961. As a result, Mao was sidelined from power but reemerged in 1966 to lead the poorly-executed cultural revolution until 1976, causing additional damage to China's already-suffering economy. Mao finally died in 1976.

Some argue that China was arguably worse than when Mao took power. In 1978, his successor, Deng Xiaoping, took the throne.

China's Economic Revolution

China has its problems, but nothing compares to how the country's economic revolution of the late 1970s propelled them from one of the poorest countries in the world to the world's second-largest economy.

Mao Zedong's death was one of the best things to happen to China's economy. His death made room for new systems of economic reforms that pulled hundreds of millions of Chinese citizens out of poverty, added decades of double-digit GDP growth to the economy, and transformed Chinese provinces into world-renowned tech hubs.

In late 1970, Mao's successor, Deng Xiaoping, had a vision to modernize China in ways only dreamt of by his predecessors. He wanted the country to experience the prosperity Western countries enjoyed using a Western-style economic model, so he established a host of land, labor, and economic reforms, eventually pulling China's economy out of its bust. This marked the beginning of China's economic revolution.

In 1978 the Chinese government opened the country's borders to international trade, which brought in business, capital, and employment. They incentivized local leaders to compete for their provinces to obtain the highest GDP growth. They allowed private companies from the outside to set up businesses in China. They also encouraged people to get jobs and start businesses.

China eventually became the largest exporter of goods and services in the world. The international community dubbed China's three-decade economic transformation as nothing short of a miracle.

Looking back to where China began to pull itself up, it's important to recognize that the government's fear of economic uncertainties was born from decades of unconscionable tragedies. It certainly doesn't justify denying their citizens opportunities to participate in crypto trading and mining, but it gives us a baseline to what China's position toward cryptocurrency is and how difficult, if not impossible, it will be for them to change that position.

Make no mistake about it, though local leaders in China maintain economic autonomy over their provinces and billion-dollar privately owned corporations are plenteous in the country, you will see that the spirit of China's old communistic ideologies still exerts influence over the government's decision-making processes. The government is still very much in control of the country and will influence the economy in any way necessary to protect its establishment.

Banning bitcoin was just the beginning. The PBOC had more extensive, broader prohibitions for crypto in China.

The Chinese Government Bans Crypto Mining

Prohibiting Chinese citizens and financial institutions from participating in crypto-related activities is one thing, but cutting cryptocurrency off from the source is a more effective way to ensure that financial independence is not reached in the country.

If you told me two years ago that China would go from being the crypto mining capital of the world to the world's most crypto mining-hostile country, I'd have thought you were crazy. This is especially true since many countries today profit from tax revenue from crypto mining. In 2021 alone, bitcoin miners generated $15.3 billion in revenue, according to The Block research.[176] I mean, what sane government wouldn't want a piece of that?

As I said at the start of the book, this book is non-technical. However, to understand why mining is so controversial in China (and other parts of the world), you must understand what mining entails and why so many people in the crypto space participate in it. So, if you don't mind, I'm going to get a little nerdy on you and give you an easy-to-understand explanation of how crypto mining works:

Mining is the process of using a computer(s) to solve a cryptographic puzzle to mint new coins (whether bitcoin, ether, dogecoin, etc.) Miners are the folks who operate the computers that do the mining, so when you hear someone say, "I'm a bitcoin miner," they own a host of computers that do the calculations to solve the puzzle. *Mining* is also used to validate and route cryptocurrency transactions, secure the network, and add new blocks to the blockchain.

Makes sense so far? Okay, let's dive a little deeper.

The goal of a miner is to be the first one to solve the puzzle because they will receive a reward in crypto.

Take bitcoin miners, for example. The current reward for successfully solving the puzzle and mining a block on the Bitcoin blockchain is 6.25 bitcoin. So, if a miner mined one block in a day, and the price of bitcoin was $50,000 on that day, then the miner would net $312,500.

That is a nice chunk of change for a not-so-hard day's work, right? So now you see why crypto mining is such a lucrative sport and why entrepreneurs all over the world flock to it.

Concerning bitcoin mining in China, before the ban, the country's open landscape and cheap electricity offered miners an ideal location for mining operations.

If you recall from the beginning of the chapter, China previously "banned" cryptocurrency in several ways. However, the first time the Chinese government considered killing bitcoin mining was in mid-2019 when the National Development and Reform Commission (NDRC)—China's economic planning faction—made plans to place bitcoin mining on a list of possible industries to eliminate from the country.[177] Oddly (very oddly), when the final list was published in September of the same year, bitcoin mining wasn't on it. It's unclear why the Chinese government removed mining from the list, but they certainly made up for it in 2021.

In May 2021, the Chinese government called for a nationwide ban on bitcoin mining, citing the ban would *"...resolutely prevent the transmission of individual risks to the social field."* The PBOC doubled down in September and re-announced its commitment to push crypto mining out of China.[178] The announcement spooked some of the major crypto exchanges like Huobi, KuCoin, and OKEx, who immediately cut ties with clients in China, while BTCChina, China's first and largest crypto exchange, completely abandoned their Bitcoin-related businesses.[179]

And if that wasn't enough, in November, the NDRC tripled down on their pledge to eradicate cryptocurrency from the country and sought measures to push out the last remaining crypto miners.[180]

The Crypto Market Became Numb

If you were tracking crypto around the April 2021 to May 2021 timeframe, you'll remember that just a few weeks before their May announcement, bitcoin reached an all-time high of $64,863.[181] Bitcoin (and other cryptos) was exploding, and China was ground zero for bitcoin mining operations. Approximately 65% of bitcoin mining took place in China.[182] Bitcoin mining operations in China all but dominated Bitcoin's global hashrate. So, when news of the mining ban circulated, you can imagine how the crypto market reacted.

The community was thrown into a frenzy and the panic selling commenced. Just hours after the initial ban was announced, bitcoin's price plummeted, right along with its hashrate. And since that horrible day in May, Bitcoin's global network hashrate continued to sink.

Okay, nerd time:

Hashrate is the amount of total combined computing power that's being used to mine a particular cryptocurrency, like bitcoin.

In other words, a hashrate captures how much mining is being done around the world.

We use the hashrate to determine the health of a blockchain network: high mining activity leads to high hashrates. The higher the hashrate, the more coins that are minted and placed into the markets. Conversely, low mining activity leads to low hashrates—the lower the hashrate, the fewer coins that are minted and placed into the markets.

The interesting thing about Bitcoin's global hashrate is that it not only determines the health of the Bitcoin network but also the health of the overall crypto market since the price of other cryptocurrencies tends to mimic the highs and lows of bitcoin.

Bitcoin's hashrate fell precipitately as more miners moved out of China; bitcoin's price naturally fell with it. This triggered a chain reaction and sent the entire crypto market into a free fall. The FUD spread fast through social media. Naysayers on Twitter, Reddit, Facebook, and YouTube heralded how China's ban would be the end of cryptocurrency and advised traders to pull their money out of the market before it was too late. And because many inexperienced crypto traders believe that everything on the internet is true, they took the FUD at face value and sold.

Bitcoin lost 55% of its value in May alone,[183] and in June, fell to its lowest price since the start of the boom in late 2020. Those of us who'd been through a few of these selloffs knew that it was only a matter of time before the bad news out of China would fizzle out and Bitcoin would bounce back, so we took advantage of the market downside and HODLed. As expected, Bitcoin began to rebound nicely in August.

What's the Real Reason China Decided to Ban Crypto Mining?

The question, however, that's been on everyone's mind was, why? Why did China kick cryptocurrency out of the country? What would they benefit from it? Were they really trying to protect the public's financial interests, or was there more to it? We may never get an accurate answer to this question, but given China's history of tyranny, one could pose a few theories:

Capital Control

Communist China's early twentieth-century ideologies still influence today's Chinese government. The essence of civil liberty suppressions from the past is still engrained in the fibers of the Chinese government almost to the point of paranoia, and it's become very apparent in the way they run the country:

- There are tens of millions of surveillance cameras set up throughout the country—some right outside the doors of citizens' residences.

- Chinese government-monitored technology and telecommunications companies have highly restricted people's access to internet content—especially those of news sites, search engines, social media apps, and messaging apps.[184]

- Chinese students at American universities are surveilled on campus and encouraged to narc on each other for openly disagreeing with Chinese policies.

- China remains the world's number one jailer of journalists for the third year in a row with 50 journalists arrested and placed behind bars in 2021 alone.[185]

The government's control over its capital is no different. The Chinese yuan is not freely-traded on foreign exchanges like other major global currencies. And unlike the U.S. dollar or the Japanese yen, which have free-floating exchange rates, China maintains strict control of its yuan and can manipulate the rates as they see fit.[186]

Cryptocurrency is non-state money. It represents true ownership and serves as a tool for financial independence regardless of the government's monetary policies. This could be one reason why the Chinese government moved at breakneck speed to drive cryptocurrency out of the country.

Let's explore a few reasons why the government disapproves of cryptocurrency in the country:

1. **China's Control over Transfers, Banks, and Spending:** The Chinese government is hell-bent on keeping the yuan active within Chinese borders. Over the decades, they've enacted policies prohibiting citizens from ditching the yuan for foreign currencies. For example, citizens are prohibited from making more than $50,000 in foreign currency a year. Likewise, the daily limit for international transfers is $50,000. (Approval is required if more than $50,000). China's growth model is reliant on these kinds of capital controls. [187]

    Here's the thing, when funds generated from savings and investments are kept within Chinese borders, those funds will be maintained in a bank. And since the Chinese government exercises meticulous control over China's banking system, they can direct funds from those banks to wherever (or whomever) they please.

The government cannot control or direct cryptocurrency the same way, so they probably would find any opportunity to expel it from the country.

2. **Monetary Policy:** As of 2020, the Chinese government made a few substantive changes to its monetary policy. In November, China's President Xi Jinping announced its long-term (and ambitious) plan to double its GDP by 2035. However, for the government to meet this goal, it must increase the country's GDP by at least 4.7% every year for the next fourteen years.[188] This will not be an easy task to complete, especially if Chinese citizens opt for cryptocurrency instead of the Yuan. As with point number one, above, it makes more sense for the government to restrict purchasing foreign currency (and cryptocurrency) so that the yuan maintains its value.

3. **Decrease Competition of the e-CNY:** I discussed the e-CNY a bit in the previous chapter, so I won't waste a lot of time going over the same details. However, I want to highlight a few ways the PBOC has used the e-CNY to assert its superiority over cryptocurrency and has attempted to satisfy the country's residual crypto cravings and make people forget that bitcoin existed.

After the Chinese government kicked cryptocurrency out of China, they spent millions on TV, internet, newspaper, and billboard ads encouraging citizens to use the e-CNY. Take the 2022 Winter Olympics, for example. In February 2022, China debuted the e-CNY to foreign athletes and visitors at the

Winter Olympics. They wanted the e-CNY to receive international attention by making guests, visitors, spectators, and athletes use the e-CNY as one of the only methods of payment.

The E-CNY to Assert Global Currency Dominance

For years the Chinese government has pushed to make the yuan overtake the U.S. dollar as the world's most dominant currency,[189] and they've been making progress. For example, in 2016, the IMF added the renminbi (yuan) to the *Special Drawing Rights* drawing basket, which is an international reserve asset created by the IMF to supplement its member countries' official reserves.[190] This has supplied liquidity to the yuan by making it accessible to international trade partners. And though the yuan now makes up only a little over 2.5% of the global reserve currency pie, the PBOC continues to make calculated moves to put the yuan on top. China might have a shot at accomplishing this using the e-CNY.

When it comes to being the global trade currency of choice, the U.S. dollar has it locked down. This is partially because the U.S. has primary influence over the Society for Worldwide Interbank Financial Telecommunications (SWIFT). SWIFT is an international payment system that enables businesses to make cross-border payments and trade internationally. SWIFT is not as independent as many foreign nations would like it to be. By regulation, SWIFT cannot service individuals, banks, or financial institutions in countries sanctioned by the U.S.[191] Sanctions, of course, can have profound effects on the sanctioned country's economy.

But with the e-CNY, China could potentially bypass the SWIFT payment system and create its own payment system that enables peer-to-peer capital flows with international partners. The government's potential to sanction-proof their economy will make room for the yuan to surpass the U.S dollar. That said, the Chinese government is unlikely to jeopardize this opportunity by allowing citizens to ditch the yuan in favor of cryptocurrency.

Environmental Concerns

The one drawback to mining is the amount of equipment required to mine popular cryptocurrencies like bitcoin. Mining in the beginning days of bitcoin could be done with a laptop sitting on your kitchen table. Things are obviously different today. The difficulty level in Bitcoin's network puzzles has increased because of the increased number of miners mining bitcoin. These days, for miners to keep up with the competition, they must stand up warehouses full of the biggest, bestest, and fastest mining computers. Some miners spend hundreds of thousands of dollars on mining equipment.

However, these powerful mining machines come with more than a heavy price tag. In addition to all the money, time, and logistics required to build mining farms, operating these heavy machineries consumes a vast amount of electricity, which can wreak havoc on a nation's environment.

In China, approximately 62% of the country's electrical power is generated by burning coal.[192] Unfortunately, this doesn't sit well with other nations because China is the world's leading emitter of greenhouse gases and mercury.[193] Out of the top five world polluters in 2019, which contributed about 60% of the world's overall pollution, China alone generated the same amount of carbon dioxide as the other four countries combined. Pollution is so bad in China that people wear masks to shield their faces from atmospheric pollution.

Though China is making efforts to reach peak carbon dioxide emissions before 2030 and carbon neutrality before 2060,[194] they have a long way to go.

The Chinese government is justified in doing whatever it takes to expel excessive coal-related activities from the country to reduce pollution. However, what if there was a way to minimize crypto mining's effect on the environment without kicking miners out of the country?

China was (and still is) the world's largest investor in renewable energy for over a decade, with $758 billion invested overall.[195] The country also has the world's largest wind and solar power capacity.[196] Let's look at the numbers. China holds 22 sizeable hydroelectric power stations distributed throughout the country[197] that, in 2020, pushed out approximately 1,350 terawatt-hours of electricity[198] to power communities, cities, and homes in the neighboring areas. China produces enough wattage to power several crypto mining rigs.

That said, there were probably plenty of opportunities for the government to work with miners and design government/provincial legislation surrounding mining with renewable energy sources.

Trash and Treasures—The Great Mining Migration

So, the crypto mining thing didn't work for China, but the story doesn't end there. China's trash became other countries' treasures. Now, other countries are in a position to reap the benefits of crypto mining (innovation, capital, jobs, recognition, a leg up on the future of finance) that China left on the table with crypto.

Shortly after the mining ban in May 2021, mining companies sought refuge in countries with cheap energy and relaxed crypto regulations.[199] Their countries of choice: the U.S., Canada, Kazakhstan, Afghanistan, and Russia. Kazakhstan, Afghanistan, and Russia particularly gave mining companies a logistical advantage due to their proximity to China. This meant miners paid less to ship equipment, and the equipment was less likely to become damaged during relocation.

Here in the U.S., Texas has become the biggest beneficiary of orphan mining companies with companies either moving their facilities to, or building mining facilities in the state. Texas has always been the ideal state for companies (and people) looking to mine crypto. Not only is it flat and spacious, but its power grid is deregulated,[200] making power providers compete for business by offering competitive energy prices. China's mining companies recognized these advantages and headed for the Lone Star state.

What also makes Texas so attractive for crypto miners is its push for renewable energy. Wind-, solar-, and hydro-generated electricity accounts for over a fifth of Texas's total electricity production.[201] That's impressive given the state's size. When looking at it in terms of land area, Texas is pushing out enough renewable energy to cover the states of New Jersey, Connecticut, Rhode Island, and Delaware. And because crypto mining's effects on the environment have been a hot-button issue for many environmentalists and lawmakers in the second half of 2021, having renewable energy as a source for mining operations supports Texas's efforts to become the largest crypto mining hub in the country.

The Global Shade Movement

The great mining migration did wonders to bring Bitcoin's hashrate back to stable levels and restore investors' confidence, but I can't say that the crypto industry is happy about how China handled the cryptocurrency issue. Not only did the abruptness of the ban and subsequent kicking out of crypto miners from the country stall the crypto market's upward moment, but the aggression with which the Chinese government libeled cryptocurrency in the press supported the one-sided narrative that cryptocurrency was bad for the world. A part of me believes that the crypto industry saw the ban coming a mile away; we just didn't know when it would happen.

I want to pose a thought-provoking question to you. Do you think it was a good thing that China banned Bitcoin sooner than later? I ask because a later Bitcoin ban—after Bitcoin and alternative cryptocurrencies became widely adopted and fully engrained into the fabric of the financial system—would have had catastrophic effects on the crypto market and the world economy.

I think we should be cautiously optimistic, though. The merits of cryptocurrency—particularly its fungibility, ability to produce financial freedom, and capacity to protect against inflation—may be enough to sway world governments to give it a place in their financial systems so that people can *legally* access and trade it. Unfortunately, I must say that crypto needs China's massive retail market to have the economic and financial impact we originally envisioned.

But China has a way of surprising us. It might be a longshot, but

I'm holding out for the Chinese government to change its stance on cryptocurrency and allow crypto trading and mining back into the country. In the meantime, we can rely on the countries that picked up the crypto mining slack to keep us afloat.

CHAPTER TEN

EL SALVADOR
President Bukele Vs. Everybody

O n September 7, 2021, President Nayib Armando Bukele gave the U.S. dollar the middle finger. On that day, El Salvador became the first country to make Bitcoin legal tender.[202] Salvadorans were now allowed to pay for goods and services (including bills and taxes) with bitcoin. The crypto-community cheered on as it was a bold leap into the country's financial revolution and a colossal achievement for the crypto industry.

However, the move appeared to have made more enemies for El Salvador than friends, and President Bukele is catching heat from all sides: 1) The international finance community believed that the president put his country in financial danger by adopting a high-ly-volatile asset as legal tender. 2) The U.S. government accused the president's cabinet of corruption. 3) And some Salvadorans felt that the president made the wrong move and resources should have been directed toward urgent matters that affected communities—like the gang violence or the water crisis.

We are going to follow El Salvador's journey to Bitcoin adoption and examine whether President's Bukele's move could be detrimental to El Salvador and to the crypto-industry.

El Salvador's Economic History

El Salvador is geographically the smallest country in Central America, but the most populated with six and a half million people living in a country the size of New Jersey.[203]

The country garnered international attention during a 12-year (1980–1992) bitter civil war that left 75,000 Salvadorans dead. The war, which set the left-wing Soviet/Cuba-backed Farabundo Martí National Liberation Front (FMLN) against the U.S.-backed El Salvadoran Armed Forces, stemmed from decades of political and military repression, ideological differences, and social inequalities.[204] The war ended in January of 1992 after both parties signed a peace agreement negotiated by the United Nations.[205] However, what followed was a slow economic recovery that was abruptly interrupted by Hurricane Mitch (1998) which killed 11,000 people[206] and an earthquake (2001) that killed 844 people and destroyed 260,000 homes and buildings.[207] The war, natural disasters, and high crime and poverty caused hundreds of thousands of refugees to flee to the U.S., Canada, and neighboring countries. Today, 20% of Salvadorans live abroad.[208]

El Salvador has seen its share of crises throughout the years, and their political and economic instability made them vulnerable to exploitation by wealthier nations. After the end of the civil war, the National Republican Alliance (ARENA) party, which was the ruling party of El Salvador at the time, initiated a series of neoliberal reforms to modernize El Salvador's economy, including replacing the Salvadoran colón with the U.S. dollar (2001).[209] The reforms led to massive privatization of the country's infrastructure and resources including electricity, telecommunications, ports, airports, water, and more.

The effects of privatization had not been good to El Salvador. Public workers turned private saw their wages slashed, labor unions that once fought for labor rights for all Salvadorans were eliminated, and workers were even laid off. As more sectors were privatized, citizens found fewer opportunities to work. The U.S. and multinational co-operations took advantage of El Salvador's privatization and flooded the country with new markets to increase profits for investors, but this reportedly did little to help the average Salvadorian looking for work or to help build a stronger economy.[210] To sum it up plainly, Salvadorans were being crushed under the weight of privatization and did not see an easy way out.

Enter, Nayib Bukele

But then came Nayib Bukele, a thirty-something technology fanatic who in 2019 ran for president. Typically seen wearing a leather jacket, denim, and riding a Yamaha motorcycle, Bukele was considered an outsider, someone who ran against the old-style political system and acknowledged the corrupt ways of the former presidents. (His campaign slogan was: "There's enough money when nobody steals.")[211]

He relied on social media to gain the favor of voters (especially young voters) rather than traditional campaigning. His presidential victory was all but a landslide. On election night, Bukele won nearly 54% of the vote with only 87% of the votes counted, which promptly secured his victory. He was the first elected president in 30 years who was not from either of the major political parties of the civil war (ARENA and FMLN).[212]

He was young, energetic, and ready to propel El Salvador into a new era. A few years after his election, a small inconsequential tourist spot on El Salvador's Pacific coast sparked a vision in him that would fundamentally change the way Salvadorans did money.

Bitcoin Beach

Hundreds of surfers visit the small, rural beach town of El Zonte, El Salvador, every year to take advantage of the high waves. Other spectators travel there for another thing: to experience the town's stores, shops, and restaurants that run exclusively on Bitcoin. They know El Zonte by its nickname, Bitcoin Beach.

Bitcoin Beach started in 2019 as a community development project to bring opportunities to its 3,000 residents.[213] Back then, El Zonte was a hotbed for gang violence and poverty and was financially secluded from the rest of the country. The people of El Zonte, especially the youth who saw no other option for survival than to join gangs, had little access to resources to progress educationally or financially. The 2019 project inspired an anonymous doner to give $100,000 of bitcoin to El Zonte under one condition: the bitcoin could not be converted into U.S. dollars but was to be used by people and merchants in everyday transactions.

The project started against the backdrop of the COVID-19 pandemic, which caused businesses in El Zonte to close, leaving families out of work and unable to pay for food and rent. Mike Peterson, one of the community project members, had an idea: he turned the $100,000 bitcoin into a monthly stipend to support families through the pandemic. Families paid for food and rent with bitcoin and grocery store owners paid employees in bitcoin.

You can probably imagine the deep appreciation people in El Zonte had for Bitcoin when the worst of the pandemic ended and the economy began to reopen. Bitcoin had become such a natural part of their lives that they continued to use it to pay for everyday items even when U.S dollars were available again. One business after another started accepting bitcoin until everything in the small town was purchasable with bitcoin. The town took on a new identity that inspired a new way of life for the people.

President Bukele saw the impact of Bitcoin on El Zonte and considered it the missing link to a new, thriving El Salvador. It inspired him to make the bold decision to make Bitcoin legal tender in El Salvador alongside of the U.S. dollar.

Bitcoin Law

President Bukele made the announcement at the Bitcoin 2021 conference in Miami on June 5, which was a fitting venue given that the conference was filled to the brim with attendees and featured members of the original Cypherpunks, billionaires, celebrities, and pro-crypto U.S. politicians. After he made the announcement, the news quickly made headlines. The Legislative Assembly of El Salvador approved the draft Bitcoin bill on June 9 [214] and it officially went into effect on September 7.

Right off the bat, the Bitcoin law does three things:

1. It gives unbanked Salvadorans quick and easy access to financial services.

2. It reduces the costs of remittance payments from outside the country.

3. It reduces reliance on the U.S. dollar and the monetary policies that accompany it.

Remittance is probably the most crucial benefit of the Bitcoin law. 20% of El Salvador's GDP comes from remittance payments. That is an astounding number and demonstrates that families (and the country) rely heavily on money transfers from outside the country. And, with El Salvador's subpar banking infrastructure, money doesn't get around as easily as in developed nations. For example, in the U.S., we can send money to friends and family members instantly with the click of a button with payment processors like PayPal, Cash App, and Venmo. In El Salvador, however, that is rarely an option. 70% of Salvadorans don't have a bank account, so collecting remittance payments can be a burdensome task.[215] Salvadorans sometimes have to travel far distances to brick-and-mortar payment services like Western Union to collect money.

Transaction fees for remittance payment can be expensive, which reduces the amount of money that would be received if fees were lower.

Bitcoin on the other hand gives families in El Salvador an avenue to receive (and send) money from any part of the world right to their mobile phones for little transaction cost.

Bukele Vs. the IMF

El Salvador could not go with Bitcoin as legal tender narrative too long without catching the attention of the broader financial community. President Bukele faced immediate criticism from several international financial intuitions, chief among them the International Monetary Fund.

The International Monetary Fund, or IMF, is a financial institution made up of 190 countries including the U.S., Europe, and other wealthy nations that oversee the international monetary system.[216] They were one of the financial institutions that emerged from the United Nations' Bretton Woods Monetary conference in New Hampshire in 1944. According to the mission statement on their website, the IMF works to achieve sustainable growth and prosperity by supporting economic policies that promote financial stability and monetary cooperation. In simple speak, they give loans to poor, struggling countries in crisis to stimulate economic growth.

The IMF had issued several warnings against El Salvador's move to make Bitcoin the country's legal tender even months before the Bitcoin law would go into effect. The IMF's head of communications made their stance on the subject clear during a June 2021 press conference, saying that El Salvador's adoption of bitcoin as legal tender raised several macroeconomic, financial, and legal issues. At the time, President Bukele requested $1.3 billion from the IMF to revitalize El Salvador's economy to include building a new airport and a Pacific train line, but talks have stalled for a myriad of reasons, one being the country's adoption of Bitcoin.[217]

El Salvador still did not have the funds to establish the technical framework required for a nationwide digital currency program. Aside from purchasing enough bitcoin to fund the entire country and keep on their balance sheet, they needed money to develop the Chivo wallet app, which citizens would use to store, send, and receive bitcoin; manufacture and station 1,500 Bitcoin ATMs throughout the country; hire auditors, cybersecurity, and customer service professionals; and pay for advertisements. President Bukele sought financial assistance from the IMF's institutional cousin, the World Bank, but his request was rejected. In June, a spokesperson for the World Bank wrote an email to the *Washington Post*[218] stating that they could not support El Salvador's implementation on Bitcoin as legal tender due to the *"environmental and transparency shortcomings."*

Some members in the crypto-community criticized the IMF, saying that they were attacking El Salvador because they were afraid that other countries would follow El Salvador's lead and adopt Bitcoin as legal tender, which—if enough countries did—could hypothetically render the IMF unnecessary. The community's claim may have some truth to it. After the president's announcement, Mexico, Panama, and Paraguay expressed interest in pushing for Bitcoin legalization and even hinted at making it legal tender.[219] [220] [221] In April 2022, the Central African Republic officially adopted Bitcoin as legal tender, making them the second country to do so. If this trend continues, then the IMF may in fact shatter (time will tell).

We will circle back to the IMF in a moment.

Bitcoin City

President Bukele's vision for Bitcoin expanded beyond making it legal tender in the country. His goal was to make El Salvador the leading country in crypto innovation and application. On November 20, 2021, at the closing ceremony of the Latin Bitcoin conference in Mizata Beach, El Salvador, the president announced plans to create the world's first Bitcoin city.[222] The city would have residential areas, commercial areas, restaurants, entertainment, its own airport, and would be constructed in the shape of a circle (like a coin). They planned to build the city near Conchagua volcano in southeast El Salvador to enable the use of geothermal energy emitted from the volcano to power Bitcoin mining farms.

It Gets Better

Following the Bitcoin City news, the president announced that the country would issue $1 billion in 10-year tokenized bonds with an initial interest payout of 6.5%.[223] The bond would be used to raise funds to buy more bitcoin and finance infrastructure for the mining farms.

Fears of bitcoin tanking aside, the government's plans for Bitcoin City and bonds were unique and well thought out. Unfortunately, they did not have a model to compare it to. Until this point, people had only talked about establishing a city-like crypto space, but no one had yet come close to what El Salvador was proposing. It was a risky move, but it was sure to greatly increase El Salvador's tourism given the increase in tourism in Bitcoin Beach.

The IMF Got Louder

The announcement of the Bitcoin bonds seemed to be a tipping point for the IMF. Their messages to El Salvador to shut down Bitcoin as legal tender got louder.

Two days after the announcement on November 22, the IMF issued its annual Article IV concluding statement to El Salvador—a report summarizing the IMF team's preliminary findings after visiting the country. With respect to Bitcoin as legal tender, the article had this to say:

"Given Bitcoin's high price volatility, its use as a legal tender entails significant risks to consumer protection, financial integrity, and financial stability. Its use also gives rise to fiscal contingent liabilities. Because of those risks, Bitcoin should not be used as a legal tender."

and

"Banking regulation should incorporate prudential safeguards such as conservative capital and liquidity requirements related to Bitcoin exposure."[224]

The IMF suggested that El Salvador narrow the scope of their Bitcoin law and implement stronger regulation and oversight for the Chivo wallet, and incorporate banking regulations. These suggestions immediately raised several concerns from users, one being that regulation and oversight of a bitcoin payment system defeat the purpose of cryptocurrency's decentralization and strip it of the financial freedom it was meant to provide. The IMF, however, did not see it this way.

They took it a step further in a January 22, 2022, press release.[225]

Though they acknowledged that El Salvador's plan to increase financial inclusion for all Salvadorans was important, they emphasized the need for "strict regulation and oversight" of Bitcoin and Chivo, and urged the president to narrow the scope of the Bitcoin law by removing Bitcoin as legal tender.

This immediately generated laughs on social media where people highlighted the IMF's ignorance of how cryptocurrency (especially Bitcoin) worked. People reminded the IMF (once again) that Bitcoin was created to be oversight-proof and that their suggestion of regulating and overseeing Bitcoin and the Chivo wallet went against the decentralized spirit of cryptocurrency.

President Bukele wasn't short on jokes in his response to the press release. He mocked the IMF by tweeting a short video clip from *The Simpsons* that showed Homer Simpson outside walking on his hands (trying to get attention from his mom and Lisa sitting on the porch) and added the caption:

🐦 *"I see you IMF, very nice."*[226]

It was a bold move given that Bitcoin had a turbulent week and its price dropped to lows not seen since July 2021.

But, President Bukele once again showed his sense of humor and replaced his Twitter profile photo with a photo of himself wearing a McDonald's hat—a nod to an old crypto joke making fun of people losing their money in a crypto crash and returning to their jobs at McDonald's.

Nevertheless, both parties are in a deadlock. The IMF continues to sound the alarm on El Salvador's Bitcoin adoption and President

Bukele is moving forward with his plans to sell Bitcoin bonds and create his Bitcoin City.

Bukele Vs. The Economy

Individual and institutional investors use a country's credit ratings to assess a country's credit worthiness—basically, the likelihood that they will default on a loan. They especially look at the country's potential for economic growth, its inflation, unemployment, national debt, and other socioeconomic and political factors. The rating can be positive, meaning that the country is ripe for economic success and should be provided whatever funds necessary to make that a reality, or it could be negative, meaning that the country has more liability than profitability. In 2021, El Salvador received negative credit ratings from all three major credit rating agencies: Standard and Poor's (S&P), Moody's Investment Service, and Fitch Group.

In mid-September, Standard and Poor's (S&P) argued that the risks associated with making bitcoin legal tender outweighed the benefits and there were immediate "negative implications" for credit. They subsequently gave El Salvador a *B-* credit rating.[227] For the sake of clarity, S&P considers any country with a rating BBB- or above to be an "investment" grade country. Conversely, they throw countries with a BB+ or lower rating (like a B-) in the "speculative" or "junk" category.

Moody's believed that El Salvador's move would limit their ability to acquire loans from the IMF and would increase financial tensions with their international partners, including the U.S. In late July, they downgraded El Salvador's rating to *Caa1*, which is equivalent to one rating notch below *B-*. They also kept the rating on a downgrade warning,[228] meaning that the country was at even more risk to be downgraded. They stated the reason below for the rating downgrade:

"In Moody's opinion, these measures reflect weakened governance in El Salvador, raising tensions with international partners—including the United States (Aaa stable)—and jeopardizing progress toward an agreement with the IMF."

Fitch Rating warned that El Salvador would have a hard time finding an insurance company to insure their bitcoin assets. Most if not all insurance companies would not be comfortable insuring bitcoin because its price was so volatile that it posed too many operational risks. [229] Additionally, bitcoin had never been put to practical everyday use as El Salvador intended to use it.

So what do all those negative credit ratings mean for El Salvador? Generally, a government's ability to acquire loans will impact investors and companies—and companies pass on the cost of borrowing to their customers. El Salvador's poor ratings will influence their ability to borrow in the future. If the country seeks to access funds in international bond markets, they typically need a good country rating. This puts perspective on the fact that President Bukele may have alienated El Salvador from the broader financial community by making bitcoin the country's legal tender.

Despite the shade and bad reviews from global financial institutions, President Bukele is determined to make Bitcoin, Bitcoin bonds, and Bitcoin City work in his country. Yet, Salvadorans' reaction to the news was not what the president hoped for.

Bukele Vs. the People

Not everyone in El Salvador loved the idea of Bitcoin as legal tender.

For one thing, Salvadorans felt that the government blindsided them with the Bitcoin law and failed to provide them adequate time to prepare for the change and training on how to use it. Days before the Bitcoin law went into effect, Salvadorans took to the streets of El Salvador's capital, San Salvador, to protest the Bitcoin law, stating that it was a move by the president to tighten his control over the people. Protestors accused President Bukele of authoritarianism and trying to impose a monetary system on people that they never signed up for.

Two, residents and business owners did not want their savings in bitcoin because of its wild price volatility. Many were afraid that if the Bitcoin market crashed (as it frequently does) then their savings would crash right along with it.

Three, though Bitcoin was accepted at many places in El Salvador prior to the law, most people did not understand it or how to use it. Average Salvadorans treated Bitcoin as an afterthought—a type of payment system that crypto-knowledgeable people used as an option—it was never a requirement. People were jolted by the sudden announcement of the law.

The roll out of the Bitcoin law was not a smooth one either. The first few months were rough on citizens and businesses. The Chivo wallet—the government-created bitcoin wallet citizens use to store and transact bitcoin payments—failed to live up to everyone's expectations. Many users reported that they could not download the wallet from the apps store. Those who downloaded the app experienced frequent network errors and crashes, disappearing bitcoin, identity exposure, inability to withdraw bitcoin from Chivo ATMs, and ATM connectivity issues. Users also complained about bitcoin payments disappearing after paying a merchant, such as when bitcoin sent to merchants never appeared in their Chivo wallets; the bitcoin somehow disappeared into the ether along the way with no mechanism to recover it. Most buyers that encountered this issue settled their debt with USD and hoped to get a refund from the government.

And probably more alarming, some Salvadorans believed the president should have focused his resources on more urgent issues that affected them daily. For example, 7% of poor households with children lack access to sanitation and 13% do not have piped water.[230] Schools were closed for 13 months following the COVID-19 pandemic, further aggravating the learning crisis.[231] In 2018, only 55.2% of children aged one to four who suffered from malnutrition received nutritional monitoring and support.[232] In 2019 and 2020, extreme poverty rose from 5.6% to 8% and poverty rose from 30.4% to 36.4%, making El Salvador the second poorest country in Central America.[233]

That's not all. The country faces other social crises including growing national debt, women's reproductive rights,[234] the government extrajudicial killings,[235] overcrowded prisons,[236] government corruption, and probably the most distressing of all, overwhelming gang violence.

Fear of death and injury is rampant in the country. With approximately 70,000 gang members in the country, communities in El Salvador are affected by violence, extortion, death threats, and forced recruitment. It held the title of the deadliest country in the Western Hemisphere in 2016, and since then has often been referred to as the *murder capital of the world.*

The country saw its bloodiest weekend of gang violence, March 25–27, 2022, with 87 gang-related homicides, 62 of which occurred in one day.[237] [238] [239] It was the deadliest weekend since the civil war and triggered a nationwide state of emergency[240] that resulted in the arrest of over 30,000 suspected gang members and terrorists.[241] Gang violence has forced tens of thousands of residents from their homes in search of safety.[242]

With problems like these plaguing El Salvador's communities, residents can't help but question the president's priorities and whether it was the best time to upend the entire financial system while residents in poor communities suffered. On the other hand, one can make the argument that there is no perfect time to start modernizing the country's financial system. El Salvador's troubled past led to repeated setbacks. The longer the government waited for conditions to be right, the more they would have to play catch up with the rest of the world.

Bukele's People Vs. the U.S. Government

In December 2021, the U.S. Treasury Department applied sanctions on some of President Bukele's officials, including his chief of staff, Martha Carolina Recinos De Bernal, according to a department press release.[243] The Treasury alleged that Recinos was the mastermind of a multi-million-dollar corruption scheme involving COVID-19 donations and contracts.

Recinos along with President Bukele's administration officials reportedly took medical aid and personal protective equipment that was donated to the government and sold it at a higher price for personal gain.[244] They also claimed that she took government-purchased food baskets that were intended for COVID-19 relief and used them to obtain votes and support for President Bukele's Nuevas Ideas Party candidates during the February 2021 municipal and legislative elections.[245]

What does this have to do with the Chivo wallet? Beginning in 2020, Recinos served as an alternate director of Empresa Transmisora de El Salvador (Transmission Company of El Salvador) (ETESAL), a state-owned electric power transmission company responsible for maintaining and expanding El Salvador's transmission system.[246] ETESAL is the majority shareholder of Chivo Sociedad Anónima de Capital Variable (S.A. de C.V.), the private company that manages the Chivo wallet.

Because Recinos was placed on two sanctions lists consisting of people with whom U.S. companies cannot do business, that could prompt the U.S. to place sanctions on Chivo S.A. de C.V., which consequently would prohibit Salvadorans living abroad from using the Chivo app and ATM to send remittance back to families in El Salvador. If that were to happen, it would devastate El Salvador's economy given that 20% of their GDP is from remittance.[247] Also, sanctions against Chivo could sour business between them and U.S. software developers. This is particularly pertinent because, at the time I am writing this, Chivo employs a U.S. company to fix the bugs in the Chivo wallet.

There were other allegations of corruption launched at the government including their use of public funds to establish Chivo S.A. de S.V (which again, is a private company) and appointing two members of the president's Nuevas Ideas Party (who are public employees) to be the owner and alternate administrator.[248] There were also questions about how Chivo S.A. de S.V came to be. The Salvadoran media alleged that Chivo S.A. de S.V used to be named *Inversiones El Salvador No. 1* (established 1999). The name was changed to *Chivo* in August 2021,[249] just one month before Bitcoin officially became legal tender. These, however, are stories to delve into at another time.

El Salvador – for the Future

So, was Bitcoin adoption in El Salvador a good move for the country and the crypto-industry? It certainly seemed that way in the short-term. In some ways, the move reinvigorated the people's hope for a better life and economy, and it brought more credibility to crypto-currency. Sure, it was a bold move and it could have been executed better, but despite all the shade thrown on the president, sometimes governments need to make drastic changes to dig their countries out of drastic situations. And El Salvador has been in a drastic situation for a long time and desperately needs a resolution.

Will the move be lucrative in the long-term, however? I think it is too soon to tell. At least for now, El Salvador's economic future is uncertain. Inflation, slow remittance revenue, bitcoin volatility, increasing public debt, and stalled debt-relief negotiation with the IMF will constrain El Salvador's medium-term economic growth prospects and extinguish any hope for them to receive capital investments from the private sector. In addition, the plethora of stern warnings, harsh criticisms, and accusations of corruption by the international finance community paint a foul picture of the president and his administration.

However, underneath all the controversy, there's at least one positive aspect that people overlook, and that is education. Since adopting bitcoin as legal tender, the country now has a rare and unique opportunity to push education in FINTEC and prepare its youth for careers of the future. These include programmers, developers, digital architects, digital asset and blockchain consultants, and digital assets legal services. Thankfully, the Salvadoran government recognized the urgent need to develop these types of training programs, and seeks to collaborate with the academic and private sectors.

I personally cannot wait to see what will happen with El Salvador in the next five years. I'm especially excited to see their progress with Bitcoin City. If all goes according to plan, we may see other countries follow El Salvador's lead and adopt Bitcoin as legal tender. I think for now, we best sit in the corner, cross our fingers, and hope for the best.

CHAPTER ELEVEN

WYOMING, THE PIONEER STATE

Wyoming has positioned itself as a crypto haven for entrepreneurs looking to establish a crypto business in the U.S. Since 2018, Wyoming has passed 24 blockchain and cryptocurrency bills ranging from crypto-friendly tax regulations and classifying crypto as *personal property* to legalizing the creation of cryptocurrency banks and Decentralized Autonomous Organizations (DAO) LLCs.[250] Additionally, Wyoming is the first state to legally codify cryptocurrency into three categories, *Virtual Currency*, *Digital Asset*, and *Security Token*.[251]

Wyoming is arguably becoming the most crypto-cordial state in the country. But why now? How are they doing it? And what kind of obstacles did they have to overcome to do it?

Before we dive into these, let me tell you what I learned about Wyoming.

Wyoming Is a Real Place

We rarely hear about the goings-on of Wyoming, whether in the news, on social media, or in everyday conversation. I shamefully admit there were periods in my life where I forgot Wyoming was a state. I'd hear it mentioned somewhere in an educational documentary and think, *"Oh yeah, Wyoming is one of ours, isn't it?"*

It's a pitiful excuse for not knowing more about it, but can you blame me? Wyoming is a quiet state, a neglected state. There are no professional sports teams, only two escalators in the entire state, and only one public four-year college, so Wyoming doesn't have a lot of headline-grabbing drama happening there. Heck, there's even an active subreddit titled *r/WyomingDoesntExist* that you can go to right now and view *recent* posts from people who legitimately question its existence.

However, now that I've taken the time to dive into the state's past, present, and where they plan to go in the future, I can confidently say that there is way more to Wyoming than meets the eyes and ears. For such a reserved state, I think you'll be surprised at how forward-leaning they are, especially regarding blockchain innovation and the cryptocurrency industry.

I can't wait to go into further detail about that, but first—since we've already established that Wyoming is unduly silent and many people probably know very little about it—I want to hit you with some interesting facts about the state so that you can get a feel for the type of innovation mongers we will discuss in this chapter.

Wyoming – the 44th State

Wyoming became the 44th state of the union on July 10, 1890.[252] Almost half of the people living in Wyoming were born there,[253] and many of them live in small ranching and farming towns or mining settlements and communities. Others live in bigger cities like Casper and Cheyenne. I think it's safe to say that Wyoming is a workers' state. Coal mining, petroleum and natural gas production, farming, and livestock production make up a significant portion of their labor force and economy.

Tourism plays a significant role in their economy too. Every year millions of people visit Wyoming to experience their state parks and historic sites: wildlife, hiking, kayaking, snowshoeing, climbing, fishing, and of course, touring of the world-famous Yellowstone National Park.

Wyoming has a low population, low crime, low pollution, low cost of living, low unemployment, no toll roads, almost no traffic, and it has the country's second-highest mean elevation—standing 6,700 feet above sea level—and its highest point stands at 13,804 feet.[254]

Taxes are not hefty in Wyoming, either. In fact, it is one of the tax-friendliest states in the country. There's no state income tax, no sales tax on prescription drugs and groceries, no inheritance or estate tax, no tax on real estate sales, and low property tax.[255] That's crazy, right? Who knew Wyoming was such an accommodating state?

Wyoming – the Pioneer State

When I thought about Wyoming in the past (*past* meaning, before I wrote this chapter), I thought of it as the state with all the forests and buffalos and nothing more. I'm ashamed of that now because I find Wyoming to be one of the nation's most underrated and fascinating states. It's a hidden gem whose historic accomplishments go so unexplored in popular culture that people walk around not knowing that many of the freedoms and perks they enjoy originated in Wyoming.

Wyoming is popularly known as the *Cowboy State* and the *Equality State*, which are both legitimate and well-deserved designations given their history of cowboys and inclusive legislation.[256] But still, it doesn't fully capture just how transcendent the state is. I think a third designation should be added to the list to do it justice—the *Pioneer State*—because unbeknownst to many, Wyoming was the first state in the country to set into motion many of the trailblazing laws, trends, and movements that make the U.S. the greatest country in the world.

For example, did you know that Wyoming was the first state to grant women the right to vote (*1869*),[257] or was the first state to elect a woman governor (*Nellie Tayloe Ross, 1925*)?[258] How about the first state to establish a national park (*Yellowstone, 1872*) or the first (and only) state to have an all-inclusive state motto (*Equal Rights*)?[259] If you're reading this book and you're an entrepreneur, Wyoming has you covered as well. They were the first state to form a Limited Liability Company (LLC) (*1977*).[260]

You're probably just as surprised as I was when I first read that. But by now, it doesn't surprise me at all. While preparing to write this chapter, I had the opportunity to study the state and its people extensively, and it confirmed some suspicions of mine.

Sure, Wyomingites are sometimes quiet and reserved, but they are also ferociously serious about their personal rights and freedoms and will not wait for the U.S. government to tell them how to do things. They are forward-thinking and creative, and they are not the kind of people who mimic everyone else's trends but set their own. I believe their self-sufficient "pioneer" mindset gives them an evolutionary edge toward innovation.

And now that they've planted seeds into the grounds of FINTECH, specifically blockchain and cryptocurrency, I am excited to see the value they will bring to the space and to their state.

Okay, before we continue, I want to give you a few more mind-blowing facts about Wyoming for good measure:

- They established the first U.S. National Forest *(Shoshone National Forest, declared by President Harrison, 1891).*[261]

- They are the largest producer of coal in the U.S.[262] and have the largest coalmine in the world *(North Antelope Rochelle Mine).*[263]

- They are the least-populated state in the U.S.

- The first J.C. Penney store opened there *(By James Cash Penney, 1902).*[264]

And Now, They're Pioneering the Crypto Industry

You probably guessed by now that the Wyoming Legislature is not afraid to take risks. The laws they set into motion—some well over 100 years ago—have changed the nation's trajectory, and now their trailblazing spirit has spilled over into the crypto industry.

I cannot even begin to talk about Wyoming and cryptocurrency without mentioning Blockchain evangelist Caitlin Long. Long is a native Wyomingite who served 22 years in corporate finance on Wall Street, working for Morgan Stanley, Credit Suisse, and Salomon Brothers.[265] She's the founder and CEO of Custodia Bank (formerly *Avanti Bank*), a new type of depository institution specializing in cryptocurrency assets, which we will discuss later.

She has an authentic love for blockchain and cryptocurrency, with her roots in Bitcoin dating back to 2012. In other words, she's a corporate beast who's achieved "OG" status in the crypto-community. A very rare breed.

Long was (and still is) essential in designing Wyoming's comprehensive blockchain and cryptocurrency legal framework. She assisted the Wyoming Legislature in drafting those 24 blockchain and cryptocurrency laws that set Wyoming on course to becoming the "Delaware of digital assets law."[266] Even more remarkably, she chairs WyoHackathon, a non-profit blockchain and cryptocurrency event at the University of Wyoming. So, if the crypto-community has dubbed SEC Commissioner Hester Peirce *#CryptoMom* then Caitlin Long should be dubbed *#CryptoAunt*.

Long talks about *attack vectors*—legal vulnerabilities in crypto businesses that state and federal regulators exploit to deny crypto businesses rights they would afford to traditional businesses.

One of those attack vectors is the relationships between banks and crypto businesses, specifically, the ways that crypto businesses can access U.S. dollars for business expenses. I want to outline some of the pitfalls that crypto businesses encounter when trying to grow, and show you how Wyoming's new crypto legislations are solving these problems.

Avoiding the LLC Trap

LLCs have become the entity of choice for cryptocurrency entrepreneurs looking to start a business. An LLC is a business structure, a legal entity separate from its owner and responsible for its own debts and obligations. Think of an LLC as a protective layer that prevents business owners from becoming personally liable for debts incurred by the business. For example, if an employee decides to take legal action against the business and sue for negligence, the LLC will be liable for any debts owed, not the actual owner. This ensures that the owner's personal assets (house, car, stocks, etcetera) go unaffected. LLCs are affordable, easy to set up and provide the members a host of tax benefits.

However, one catch to having an LLC is that owners must establish a business bank account to fulfill certain tax obligations, be recognized as a legit entity by state and federal regulators, and acquire enough funds to grow their businesses. This is where crypto entrepreneurs fall into the *LLC trap*. Payroll tax obligations like local, state, and federal income taxes, social security, Medicare, unemployment, etcetera are paid through a business bank account. Failure to comply could lead to some nasty fees and penalties. And that's not all; crypto businesses may also need credit to cover other business expenses like supplies, tech services, legal services, transportation, office space or buildings, or cash to hire more employees. Unfortunately, these items and services cannot be paid for with cryptocurrency (yet).

States Know this and Try to Oppose It

Knowing this, some states purposely find avenues for excluding crypto startups (either legally or financially) from establishing LLCs in their states.

In chapter six, you read about federal regulators' offensive posture toward cryptocurrency and the many uncertainties in federal cryptocurrency regulations. Unfortunately, this caused laws toward cryptocurrency to vary significantly by state—some imposing more tyrannical rules on cryptocurrency than others. While states like Wyoming, Florida, and Texas embody favorable crypto regulations for businesses, states like New York impose famously hostile crypto regulations that stir contentions and freeze out companies that cannot afford to survive.

For example, in New York, entrepreneurs seeking to engage in virtual currency business activities (transmitting, storing, holding, custodying, issuing) must apply for a BitLicense,[267] which is a regulatory standard created by the New York Department of Financial Services (DFS) in 2015 to shield consumers from crypto scams and prevent crypto-related illicit money laundering activities. The requirements for the license, however, are quite controversial. For starters, as of 2022, the BitLicense application fee is $5,000, which is 25 times the cost of applying for an LLC in New York and 33 times the cost of securing a New York business license.[268]

Secondly, applicants must establish a surety bond—a three-party insurance agreement between the business, the state, and the consumer—which costs a minimum of $500,000. Third, DFS BitLicense requirements suspiciously resemble the hiring process used by the CIA to vet their future batch of super killer agents. Applicants are required to submit financial statements, bank account, credit reports, FBI criminal background investigations, a meticulously detailed business plan, good standing certificate, company staffing, and internal policies, flow of funds structure, organizational chart/description, photo IDs, financial statements, resumes or curriculum vitae, fingerprints,[269] and probably some more stuff they don't advertise. Even the name of the applicant may be subject to superintendent approval.[270] Mind you, businesses must show that they can satisfy these requirements before the DFS even considers issuing a BitLicense. And if by some miracle, a crypto business can fork up this kind of cash (and allow such invasive reporting requirements), the DFS can still decline an application if they choose to do so.

Though I applaud the city for practicing due diligence by its residents, the costs for the BitLicense alone are far above what a small startup or even a medium-sized company can afford.

Relatedly, the DFS has been accused of attempting to control the crypto market by creating license requirements that favor only specific institutions, and it's apparent in the number of BitLicenses they've issued. Since 2015, the DFS has issued only 30 BitLicenses, most of which were to wealthy financial enterprises and brokerages, including PayPal, Sofi, Robinhood, and Fidelity.[271] Several crypto exchanges, including Bitfinex and Kraken, refused to comply with BitLicense standards and stopped servicing New York entirely. Crypto exchanges left New York so often that the crypto-industry coined a term for it—the *Mass Bitcoin Exodus.*

192 • THE DEVIL MADE CRYPTO

Although no state or federal law requires business owners to establish a separate bank account for their LLC, having one significantly reduces governmental scrutiny on businesses and minimizes avenues through which federal regulators can attack businesses or business-like entities like a crypto business.

Wyoming Launched a Counterattack

The Wyoming legislature, however, has already anticipated these obstacles and structured its legal framework to make it possible for entrepreneurs, regardless of their state of residence, to set up crypto LLCs in Wyoming and benefit from their crypto-friendly laws.[272] The most apparent benefit that sets Wyoming apart from competing states is tax benefits.

Let's not pretend that some states (New York and Connecticut) don't tax the crap out of people and businesses. It can be annoying and quite disheartening at the same time. The good news is that under Wyoming law, crypto businesses that are physically located or domiciled in Wyoming are exempt from personal income tax, corporate income tax, sales tax, franchise taxes, and gross-receipts taxes.[273]

If you think about it, most crypto startups make little to no money in the first two or three years and are probably operating on a strict budget. Coinbase, a now multi-billion-dollar crypto trading platform, was started in a two-bedroom apartment by two guys who met on Reddit in 2012. Startups in these positions stand to benefit the most from this law because 1) They won't panic about not having enough money to cover the spread when tax time arrives. 2) The money they save on taxes can be thrown back into the business to pay for supplies, additional hardware and software, additional employees, and whatever else they will need to grow the business. 3) Wyoming is one of only four states (for now) that will allow a business to (directly) form an anonymous LLC.

I had no idea that something like this even existed, given that most states can be annoyingly invasive in matters involving business and finance. Under Wyoming's law, businesses do not need to disclose the names of members and employees to the public, which is crucial to privacy protection. Identity thieves and hackers are pretty resourceful, right? There are dozens of websites that bad actors use to access detailed corporate records and private information about business owners and employees; some charge very little money. If you've been in the crypto space for any time, you already know that account hacks remain the number one threat to traders, crypto exchanges, and DAOs. Denying the public names, addresses, phone numbers, and other identifying information about the business and its employees closes the avenues bad actors use to access employees' and customers' wallets.

Last but certainly not least, in 2018, the Wyoming Legislature amended the Money Transmitters Act to exempt cryptocurrency.[274] This means that crypto startups and businesses that buy, sell, or hold cryptocurrency on behalf of customers won't have to shell out a lot of money to get their business going. They're not required to pay the money transmitter license fee (which is not to exceed $3,000), obtain a surety bond, or initially have a net worth of $25,000.[275] It takes the pressure off crypto exchange startups and allows them to focus their capital and energies on innovating their businesses.

Crypto Banks

Yes, cryptocurrency banks are finally a thing! In 2019, Long, along with other members of the Wyoming Blockchain Task Force (which evolved into Wyoming's permanent *Select Committee on Blockchain, Financial Technology and Digital Innovation Technology*[276]), spearheaded a Wyoming House bill that created a new kind of financial institution called *Special Purpose Depository Institutions (SPDI)* (pronounced, *Speedy*).[277] A SPDI is a specialty bank that acts as a digital currency custodian.[278] It's a crypto bank, basically. They plan to do something that's never been done in the crypto industry—safely connect crypto businesses with the federal financial system in a way that both benefits the business and complies with U.S. federal banking regulatory standards.

You've read in the previous section that, although crypto was created to disrupt the traditional banking industry, there are things that crypto businesses cannot yet accomplish using cryptocurrency (payroll taxes and business expenses). Therefore, crypto businesses need access to U.S. dollars to expand and survive. SPDI will be that bridge that enables dollars to flow to businesses without interruption.

Apart from providing crypto custody to clients, SPDIs have peculiarities that distinguish them from traditional banks:

1. *Under Wyoming state law, SPDIs must maintain 100% of their customer's fiat demand deposits and cannot use them to make loans, while commercial banks hold only a fractional percentage (fractional reserve) of customers' fiat deposits and can lend the rest to loan seekers.*

2. *SPIDs operate under the regulatory oversight of the Wyoming Division of Banking, whose regulations are crypto-friendly, while the Federal Deposit Insurance Corporation (FDIC), which regulates commercial banks and whose regulatory framework is stringent and inflexible.*

3. *Deposits held in crypto banks are not insured by the FDIC like commercial banks, therefore, SPDI will be deemed "high-risk."*

Two banks are leading the race to become the nation's first federally approved SPDIs, Kraken Bank and Custodia Bank. (If the name *Kraken* sounds familiar, it's the same San Francisco-based company that operates the Kraken crypto exchange). The Wyoming State Banking Board awarded Custodia and Kraken charters to establish SPDIs in the state in late 2020.[279] Their next step will be to receive master accounts from the Federal Reserve, which will allow them access to the Fedwire payment system to send or receive payments on behalf of a business, organization, corporation, or other entity.

Born from Desperation

Crypto banks are an immensely prolific concept, far ahead of its time, and well beyond what the crypto community anticipated this early in the game. I wish I could sit here and say that this innovation was spawned from sheer ambition and would have little to no effect on the crypto industry if it were to fail, but that's just not the case. Crypto banks were born out of desperation.

The crypto bull market of 2017 spurred the era of the crypto startup. Dozens of cryptocurrency companies, including crypto exchanges and mining companies, sprang up all over the world in just a short period. For example, the world's largest crypto exchange, Binance, got its start by raising $15 million in an ICO in June 2017.[280]

Likewise, U.S.-based Protocol Labs raised a record-breaking $257 million from its file-sharing coin, Filecoin (FIL), ICO in the fall of 2017.[281] These are only a few examples of crypto startups birthed in the bull market fires, but unfortunately, not every startup got the same love. The 2017 bull market also spurred the era of the *bank shade* where banks found any and every reason to stop account holders from participating in cryptocurrency-related activities including freezing, closing, or threatening to close traders' accounts.

Before 2017, banks didn't give much credit to cryptocurrency because it had not received much public notoriety. Sure, cryptocurrency was gaining popularity, and people traded it, but transactions went largely unnoticed by banks. To them, it was as much of a threat to the financial system as buying an item from Amazon or eBay.

However, federal regulators were blindsided when the bull market started to gain momentum. Its rise was so sudden and meteoric that federal regulators responded in knee-jerk fashion. Federal regulators quickly reminded financial institutions of cryptocurrency's association with criminal activity like money laundering. The reminder naturally spooked banks into freezing, prohibiting, closing, or threatening to close accounts belonging to crypto traders.[282]

Between 2017 and 2018, testimonies of how banks and credit unions sent letters to individuals informing them that their bank accounts had been closed for crypto-related activities flooded crypto news channels and social media outlets.[283] [284] [285] Likewise, federal regulators regarded banks that serviced crypto businesses and startups (providing business bank accounts, loans, credit cards, etcetera) as "high risk," resulting in commercial banks cutting ties with crypto-businesses. And this, of course, caused many crypto businesses to shut down.[286]

SPDIs are set to provide crypto businesses and startups access to business accounts without the prospect of being shut down for frivolous reasons. That, in addition to sending and receiving fiat payments on behalf of crypto businesses, will give them the juice they need to operate with peace of mind and comply with U.S. tax laws.

Federal Opposition To SPDIs

The road to establishing crypto banks, like most endeavors in cryptocurrency, wasn't an easy one. Both Long and Kraken CEO Jesse Powell hoped to begin custodying crypto in early 2021 but getting those master accounts approved turned out to be more of a challenge than they expected. The banks caught some shade from the Community Depository Institutions Advisory Council (CDIAC), a cabal of representatives from commercial banks, thrift institutions, and credit unions that advises the Federal Reserve banks on the economy, lending conditions, and other issues of interest to depository institutions.[287] [288]

During a meeting with Federal Reserve bank officials in November 2021, the council outright declared that SPDIs were seeking to "avoid traditional financial institution regulation" and that if the Federal Reserve grants them access to the federal payment system, they will introduce "increased risk to the financial system."[289] These harsh, damaging words may have influenced the Federal Reserve's decision regarding SPDIs master accounts.

On January 11, 2022, Federal Reserve Chairman, Jerome Powell, told members of the Senate Committee on Banking, Housing, and Urban Affairs that there are "good arguments" for granting master accounts to Kraken and Custodia, but they will take their time doing so because: "We start granting these, there will be a couple hundred of them pretty quickly and we have to think about the broader safety and soundness implications."[290]

Right away, we see a problem. Both Custodia and Kraken filed for their master account shortly after being granted Wyoming bank charters in October 2020. That was a good while earlier. Since then, the Federal Reserve Board in Washington, D.C., to my knowledge without giving any warning, released "new guidance" to applicants applying for master accounts, making account requirements more obscure and open to interpretation. This undoubtedly lengthened the application processing time for Custodia and Kraken.

It's difficult to say what the CDIAC's true motives were behind their initial comments about SPDIs, but, interestingly, nearly a quarter of the top 100 banks (by assets under management) are currently building cryptocurrency custody solutions of their own or investing in the companies that provide them[291]—placing them in direct competition with SPDIs. Their comments didn't seem to faze Long, though. Remember, she is a native Wyomingite. Shade bounces off her like rain bounces off a raincoat. After the CDIAC's meeting transcripts were released, she confidently addressed a letter to the team at CoinDesk to reaffirm her optimism that the Federal Reserve will approve master accounts for SPDIs.[292] In all honesty, if anyone can get it done, #CryptoAunt can. So now, we wait.

DAO Is the New Black

On November 11, 2021, a group of crypto enthusiasts came together to raise money to purchase one of 13 first printings of the United States constitution at an auction.[293] They called themselves the Constitution DAO. The DAO told the crypto-community of their plan and in only three days, raised $40 million. Unfortunately, Constitution DAO lost the auction to Citadel founder Ken Griffin, but they accomplished something even more valuable: the absurdity of how quickly the DAO formed and raised capital brought awareness to the concept of the DAO as a new, workable type of business structure.

What Exactly Is A DAO?

A DAO (Decentralized Autonomous Organization) is a new type of business structure where ownership and decision-making authority are distributed across a group of people.[294] Think of a DAO as an internet-based business or entity wholly owned and operated by its members. It runs on a blockchain, and its bylaws and rules are programmed into a smart contract.

DAOs were created as an alternative to centralized organizations like companies or corporations to reduce fraud, waste and abuse, corruption, and single points of failure. And unlike traditional business structures, where the CEO and board of directors decide the future of the business, a DAO has no central authority and relies on its members to determine the future of the DAO.

Wyoming DAO Legislation

DAOs have been the talk of the crypto-community in recent years, with several DAO-related projects forming to either solve a problem in the crypto space or create something completely new and unimagined.

Wyoming certainly saw the potential of DAOs. On July 1, 2021, the Wyoming Legislature passed the *Wyoming Decentralized Autonomous Organization Supplement*, which was added to Wyoming's Limited Liability Company Act to formally recognize DAOs as LLCs in the state.[295]

Wyoming is trying to solve a problem that the overwhelming majority of DAOs encounter after they are formed, the problem of *General Partnership* and *Unlimited Liability*.

Solving Problems – General Partnership and Unlimited Liability

When DAOs are formed, they often form as a type of unincorporated association or foundation like the Badger Dao Foundation or the APE Foundation. Legally speaking, these types of associations will be treated as general partnerships when disputes arise, and with general partnerships, unlimited liability automatically applies. There are advantages to general partnerships: they are inexpensive, flexible, and easy to form, but their downside is their unlimited liability. Having unlimited liability means that every organization member is personally responsible for the entire debt incurred by the organization, even if they had nothing to do with creating it, including lawsuits.

Attorneys who coauthored the Wyoming Decentralized Autonomous Organization Supplement recognized that many DAOs welcomed legal structures for this reason. Nobody wants to work in an office surrounded by drama kings/queens, right? Likewise, DAO members and investors don't want to be part of a DAO full of drama kings/queens or open to external lawsuits without having some form of liability protection. So, Wyoming lawmakers asked themselves which legal entity would be best suited to offer DAOs the flexibility, benefits, and liability protection they desired. The obvious choice was the LLC.

Solving Problems – Wyoming's Chancery Court to Solve DAO Disputes

So, what does a DAO dispute look like? They're not pretty. There's a lot of finger-pointing and "yo mama"-ing on chat forums, and sometimes actions by the DAO cause its members to lose money. One of the more noteworthy examples of a DAO dispute is *The DAO Hack Of 2016*.

Genesis DOA Hack of 2016

The first DAO created, an Ethereum-based venture capitalist fund conveniently named The DAO, suffered an attack in June 2016 where a hacker stole approximately $60 million of Ethereum coins. (To avoid confusion, I will refer to The DAO as the Genesis DAO from here on). Weeks before the hack, Cornell computer scientist Emin Gün Sirer warned the Genesis DAO of a vulnerability in their smart contract that potential hackers could exploit to siphon funds from the contract. However, before the DAO could act against the hacker, $60 million was stolen. To mitigate the hack and restore funds to the smart contract, Genesis DAO members proposed a blockchain hard fork, which would create a parallel Ethereum blockchain where the smart contract's vulnerability was patched, and no coins were ever stolen, essentially creating a history where the hack never happened.

So, in DAO-like fashion, the DAO sent the proposal up for a vote. This was when the dispute arose. DAO members in favor of the fork wanted their stolen funds restored. Members against the fork believed that the fork would permanently damage investors' confidence in Ethereum and ruin Ethereum's chance of becoming the world's premier blockchain. This proposal became a huge controversy in the Ethereum community because if the vote resulted in negative sentiments toward Ethereum, tens of thousands of Ethereum users would risk a massive selloff and losing their invested funds. It would also call into question the feasibility of DAOs as an effective governing model and the safety of smart contracts.

The vote resulted in 87% favoring the hard fork, so Ethereum split in two. But, as you can see today, the Ethereum blockchain did not collapse: it's going stronger than ever. Unexpectedly, however, members that voted against the hard fork remained active in the original Ethereum blockchain and kept it growing, even until this day. This blockchain would be later called (yep, you guessed it) Ethereum Classic.

The Genesis DAO hack is one of few examples of a DAO dispute where the outcome would not only affect the DAO internally but affect an entire blockchain ecosystem. In December 2021, the Wyoming Legislature created a chancery court. The court was created to "expeditiously resolve disputes related to commercial, business, trust, and similar issues through non-jury trials, alternative dispute resolution methods, and limited motions practice."[296] The legislature has made it clear that the chancery court extends to DAO LLCs to settle un-settleable disputes. So, if a situation like the Genesis DAO dispute ever happened again, provided that the DAO in question is an LLC in Wyoming, it may be settled in the chancery court.

As of April 2022, there are 287 active DAO LLCs registered in Wyoming.[297]

Only In Wyoming

Okay, this chapter is already long, and I have to go to bed, so I'll leave you with this; don't make the same mistake I did, don't minimize something because it is not as glamourous as you think it should be.

I regarded Wyoming as a fly-over state and never bothered to study their contributions to the crypto-industry and the nation. However, I'm a fan now that I've immersed myself in their business, technology, and legal cultures. I'm almost sure that Wyoming is the only state that would ever consider passing a suite of crypto bills so early in the game—and we've barely scratched the surface of their plans to transform the state into the crypto mecca of the U.S. I did not get a chance to talk about their other ventures, like their mission to produce 5% of the U.S. Bitcoin mining hashrate by 2024 or the stablecoin bill that would make them the first state to offer its own cryptocurrency asset.[298]

As you can imagine, these bold moves can either end in trailblazing successes or in horrid crashes 'n' burns. But that's why the crypto-industry loves Wyoming, right? It takes risks that few other states will. Given Wyoming's track record for making unpopular trends a hit, it is probably one of only a few states that could make this work.

WHERE DO WE GO FROM HERE?

Will we ever live in a world where crypto is free of shade? At this juncture, it's too soon to say. Though some governments openly support participation in crypto-related activities, others attempt to curtail citizens' use of cryptocurrency to maintain control. They do this in a few ways: Some governments promulgate false narratives such as cryptocurrency is bad for the environment or only criminals use cryptocurrency; some governments have created or plan to create CBDCs to lure interest away from cryptocurrency and toward state-mandated money; and some governments, like China's, have full-out banned crypto activities and mining.

Outside of the need to maintain control, sometimes federal regulators and lawmakers perceive cryptocurrency as a threat because they often misunderstand what cryptocurrency is, how it works, and what its intent is. There are plenty of reasons for this, but the ones I hear the most are misinformation or disinformation from biased and uninformed sources, federal officials' outdated linear mindsets, failure to understand the technical jargon associated with cryptocurrency, and—like some of us in the past—they don't know where to go to learn about cryptocurrency.

Even we, the crypto-community, can throw shade on cryptocurrency under the right conditions. Falling victim to a scam is one of those conditions. Regardless of the favorable impact cryptocurrency has on society or one's life, as long as scams and unethical practices like pump-and-dumps, rug pulls, front running, and market manipulation exist, it will bring negative attention and wrath to the overall crypto industry.

How Can We Block the Shade and Save Crypto's Future?

There are a few ways, but they are not one-and-done solutions—they will take continuous work. I've identified three areas that, if executed properly, can reduce the amount and frequency of opposition toward cryptocurrency. These are *Education*, *Clarity*, and *Incorporation*.

Education

Education is probably the most crucial area to focus on. Federal regulators and lawmakers should seek education on the fundamentals of blockchain and cryptocurrency. This should include at least a clear explanation of the two primary types of consensus mechanisms (proof-of-work and proof-of-stake), the fundamentals of decentralization and P2P transactions, crypto wallets, how smart contracts and DeFi works, and DAOs. Knowing these things should help them better navigate the crypto landscape and will come in handy when drafting crypto legislation.

Also necessary, they should seek education on cryptocurrency's emotional and social-cultural impact on the crypto-community, including its entrepreneurs, developers, professionals, enthusiasts, and spectators. Most lawmakers buy into the narrative that criminals primarily use cryptocurrency to launder money and hide their illegal transactions. Though we know this is not a fact, lawmakers must understand that 1) the crypto-community dislikes crypto criminals as much as they do, and 2) the number of people who use cryptocurrency for nefarious purposes is minuscule compared to the number of legitimate users who want the technology to succeed. They may be more open to compromise on crypto legislation if they understand this.

The one drawback is that lawmakers may not willingly pursue education in cryptocurrency and blockchain. This is understandable because lawmakers juggle many responsibilities and can be very busy. Therefore, it is up to us, the crypto-community, to put as much good educational material in front of their eyes as possible.

In addition, the crypto community must better educate new or inexperienced crypto investors on how to spot crypto scams. Crypto scams hurt not only their victims but also the entire crypto industry, despite how small a scam might be. It tarnishes the industry's reputation, causes investors to become bitter toward the crypto space, and gives federal regulators and lawmakers ammo to push policy restricting crypto-related activities. I recommend creating more on-line resources devoted to spotting crypto scams and pointing new and existing investors toward them. I recommend that experienced crypto traders frequently promulgate detailed scam-spotting tips and techniques on social media and that those tips and techniques be updated as scams become more sophisticated.

Clarity

In America, there is still the issue of the lack of crypto regulation and the question of whether all cryptocurrencies should be classified as securities.

The lack of regulatory clarity causes confusion, frustration, and inno-vative timidity for crypto entrepreneurs and developers. Those that want to do right in the eyes of the government have no clear guard-rails to protect them from prosecution.

In the chapter on crypto regulation, you learned that not all cryp-tocurrencies should be classified as securities because many do not meet the criteria for an *Investment Contract* under the Howey test. Additionally, because cryptocurrency has unique specifications and functionalities, shoe-horning its existing federal regulatory financial frameworks will not work.

The good news is that, as you've read in previous chapters, President Biden issued an Executive Order directing federal regulators to study the crypto industry and develop relevant policy. This would be the perfect opportunity for regulators and leaders in the crypto-com-munity to come together and design a framework unique for digi-tal assets.

With that said, new crypto regulation should not be reactive but responsive—given careful thought, consideration, and feedback from the crypto community. Overly restrictive regulations may discourage innovation, drive innovation out of the country, and discourage innovation from entering the country. Conversely, overly vague regulations may give too much room for interpretation, which could later be manipulated. The world needs sound and specific regulations to steer the community in a productive direction.

Incorporation

Contrary to popular belief, cryptocurrency does not have to compete with the current financial system. I believe that both systems can coexist in a mutually beneficial way and benefit the people they serve. We saw an example of this in our chapter on Wyoming. Wyoming-based SPDIs Custodia and Kraken bank will act as intermediaries between crypto businesses looking to secure business bank accounts and the federal payment system. They plan to do this by connecting directly to the Fedwire payment system—a system that, until this point, only serviced traditional banks. With direct access to the federal payment systems, they can settle U.S. dollar payments in cryptocurrency transactions in real-time, reduce costs and transaction delays, and reduce counterparty risk. This is only one example of how both systems can coexist.

However, it is essential to note that, though both systems can coexist, nothing will be accomplished until federal regulators, financial institutions, and leaders in the crypto industry come together to hash out details and work out a compromise.

One thing is certain: discouraging participation in crypto-related activities will harm advancements in the space and put one's nation at a competitive disadvantage. Leaders should do their best to retain the innovation, nurture it, and make it work for the good of the country.

Final Thoughts

Some governments will not forfeit financial control of their countries despite the number of solutions to cryptocurrency presented to them. Likewise, some individuals will never acknowledge cryptocurrency or accept it despite how promising it is for the world. Unfortunately, there's not a whole lot that can be done about them, but we shouldn't let that stop us. We should continue to evolve as a community.

Cryptocurrency is here to stay. And as the crypto-community, we must be good stewards of the coin and stay the course Satoshi started us on. Let the world know that it's not about competition but inclusion. There will be opposition, but if we provide each other cover from the shade, we can steer the world into a better tomorrow.

MY CRYPTO STORY

veryone has a story about their first time trading crypto. Some of you probably felt excited, for others it felt weird, and for another population of traders—those who got in on bitcoin in the early days—it was super complicated. I guess you can say that my crypto story started in 2018 after the ICO boom of 2017. I watched CNBC and Yahoo! Finance religiously, and I'd hear the anchors drop the word *crypto* every now and then. I didn't think much of it initially, but the term became more frequently used, *crypto, cryptocurrency, bitcoin*.

It started to annoy me. I felt like I was missing out on something, so I figured I should see what this crypto is all about.

I loosely followed the crypto market and did some light research on projects like Ethereum, Tezos, and VeChain but wasn't ready to start trading them. I had started a new position at work and spent most of my time on the road traveling across the states. I mean, crypto seemed fun, but I had to set it to the side because I simply did not have time for it. And that was that.

So I thought…

But it's funny how ideas buried in your subconscious reemerge when you least expect them.

Life Before Crypto

I had very little exposure to technology growing up. In the Midwest, subjects like science, education, sociology, and the arts ruled in our schools, so that was what most of us focused on. The little exposure I had was from my older brother. We were good at breaking things around the house, so we got good at fixing them. He taught me how to splice copper cords, make TV antennas from metal wire, fix damaged cassette and VCR tapes, repair damaged CDs and DVDs, recharge dead batteries, and restore damaged Nintendo game cartridges. But that was about the extent of my MacGyvering. Although I didn't care so much about technical skills, I'd always had an affinity for money. I think I was born with it. When I was eight years old, I'd grab a sheet of printer paper, lie on the living room floor and draw outlines of five-dollar bills and then coloring them with a green crayon. The funny thing was, I had every inten-tion of taking those five-dollar bills to the grocery store across the street from our place and buying a pop and some potato chips!

I started investing as a teen. I got a decent job right out of high school, and it gave us the option of setting aside some of our month-ly payments in a mutual fund. I didn't know much about investing then, but I learned real fast. It seemed like a no-brainer at the time—you put the money in, the money goes up, you take the money out, then repeat. I studied all the plans in the mutual fund booklet they gave us, and I looked up the investment terms in a Webster's dictionary that I kept around my place (there was no such thing as online dictionaries back then). I don't know what it was, but it clicked instantly.

I learned the ins and outs of stock trading when I got to college. The internet started to get popular then, so I had a few online resources at my disposal. I opened a Scottrade account and went to work. I took extra classes in college so I could graduate early (that's how much I didn't want to be in college), but I found time to make trades between classes. I learned how to navigate the market mainly through trial and error instead of formal education. In some ways, I think it made me a better trader.

After college, I was hired by a financial firm in Chicago where I learned the professional side of stock trading and financial consulting. I loved every minute of it. But then, years later, 2008 happened, and my dream of climbing that corporate ladder died.

2010 – I Said "No" to Bitcoin

I first heard about Bitcoin in 2010. I had just returned from a very unpleasant year-long vacation in the Middle East and one of my coworkers, Jalen, said that he had gotten into Bitcoin and that I should start investing in it. Being the closed-minded nonconforming guy I was back then, I let out a "pshh!" and rudely told him that I wasn't putting my cash in some fake internet money you can't do anything with.

Well, he didn't react the way I expected him to. If you knew Jalen, you'd know that he doesn't take too kindly to rudeness. He was tall, wore a military-style buzz cut and scruffy grey beard, never smiled, never made eye contact, never spoke about himself or things he was interested in, and had no problem telling you how he felt about you. He was just one of those guys you wouldn't ask for a favor.

So instead of getting in my face, he smirked, looked me in the eye and said with a calm yet confident voice, "It's working for me. I'm making money from it." It was creepy. There was nothing strange about what he said, but it was strange that *he* was saying it. Jalen didn't say things like this unless there was some definitive truth about it.

Now, that got me curious about Bitcoin, and from that point on, when the topic of Bitcoin came up I'd think about what he said to me and how there must be some truth to it.

2019 – Crypto Found Me Again

In 2019, the crypto curiosity buried in my subconscious caught up to me, and one day I woke up with an overwhelming desire to invest in crypto. At that time, no one, not even long-time equity market investors in my circle, knew how to invest in crypto, so I resorted to the internet. I found a few U.S.-based crypto exchanges with weird names like *Abra*, *Binance.US*, and *Coinbase*, and went to work.

Learning about how to invest in crypto was a huge pain in my brain. Sure, Abra and Binance.US made the process of trading crypto easy, but it was the quirks of cryptocurrency that threw me for a loop: *You mean I own my own crypto, and if I lose my seed phrase or transfer some to the wrong address, no bank or company can recover it?*

I can buy a fraction of a coin instead of buying the whole thing?

It only cost fractions of a cent to send cryptocurrency across the world?

Who has ever heard of such a thing?

Once I got past myself, I pulled the trigger and started moving money into the crypto market.

My portfolio saw some gains, but what fascinated me was how volatile the crypto market was compared to the stock market. Ethereum would be trading at $150 one day and $120 the next day. Tezos would be trading at $1.15 one week and then $1.75 the next. At least to me, they seemed like crazy price swings.

Trading slowed significantly during the crypto winter of 2018–2019. Prices mostly fell to all-time lows. Even today, it's hard to believe that I personally witnessed Ethereum trading at only $50 and Bitcoin trading under $3,000. If I only knew its potential back then.

Though we were in the middle of a crypto winter, for me, cryptocurrency was fresh and exciting innovation that traded at dirt-cheap prices—I noticed its potential to rise tens of thousands of percent. In my bones, I felt that it was only a matter of time before people recognized its potential and the crypto market would blow up, so the best strategy was to buy, HODL, and wait for that day to come.

Crypto 2021

Crypto investors, including myself, who survived the crypto winter of 2018–2019 finally saw the fruits of our labor at the end of 2020 and in 2021. What went from a barely-billion-dollar crypto market soared to over two trillion dollars. The market grew tens of percent daily, and people made money. Cryptos I bought at .0002¢ in 2019 rose to 20¢. NFTs I bought at $20 in 2020 rose to $300. Bitcoin blew past $60,000 and Ethereum hit $4,000. Every sector saw gains—payments, DeFi, NFTs, predictions, you name it. New crypto projects emerged daily. Cryptocurrency was blowing up everywhere and there was no way that you could lose.

What really pulled my heart deeper into crypto was the boom's effect on the crypto community. Traders, content creators, developers, and enthusiasts that faced years of opposition and adversity watched as everything they fought for unfolded in real-time. Institutional investors jumped into the space, crypto news sites and blockchain education programs were created, blockchain organizations sprang up all around the world, mining companies stood up, and corporations and companies established blockchain research teams. Cities and even entire countries hosted massive crypto conferences. It was like the world finally awoke to crypto.

I Wanted More than Just Money

People were making money from cryptocurrency, but I wanted more than just money; I wanted to learn more about this mysterious disruptive technology called *Blockchain*. Mind you, I liked technology back then but did not have a techy-type mind. Apart from my iPhone, I didn't care what technology was out there or how it worked. I felt that learning about that stuff only applied to engineers, scientists, and the folks in Silicon Valley, not to regular guys like me. Nevertheless, I pushed past all the self-doubt and gave it a try.

First, I needed to find a resource, so I did what any high-speed 30-something-year-old American boy would do to learn a new skill, I watched YouTube videos. The first video I watched was called, *What Is Blockchain?* made by a guy who'd been in the space for several years. After about 90 seconds into the video, my mind was on tilt.

I had no clue what this guy was talking about. He was using words I had never heard before, like consensus protocols, hashing algorithms, technology stacks, and so on. It sounded like Greek to me, and at that moment, I began to waver. What started as excitement turned to dread, and I asked myself, *"Do I really want to commit time and brain power to learn about blockchain? Is it even worth the effort?"*

I knew that if I wanted to learn about the ins-and-outs of blockchain, I needed something more than excitement to fuel my journey. So, instead of diving nose-first into learning the technical components of blockchain, I started with a thousand-foot overview of the blockchain space and what it could offer the world.

What Worked for Me

Everyone has a distinct way of learning new things. I didn't have the mind of a tech genius, but one thing that I did understand was the order of events. My years working on various team projects in the professional world taught me that for any project to be successful, everyone must do their assigned jobs at the right time and in the correct order—and this is generally how proof-of-work blockchains function. Look at the process that transactions must endure on the Bitcoin blockchain, for example:

1. Users initiate transactions from their Bitcoin wallet.

2. Transactions are broadcast to the miners on the network.

3. Miner authenticates the senders' digital signature, wallet address, and funds, and ensures the transaction is not a duplicate.

4. Transactions are stored in the unconfirmed transactions memory pool.

5. Miners pull transactions from the memory pool and stuff them inside candidate blocks.

6. Miners compete for the opportunity to add their block of transactions to the chain by solving a cryptographic puzzle.

7. The first miner that solves the puzzle broadcasts the answer to the other miners.

8. At least 51% or more of the miners confirm that the answer is correct.

9. The winning miner confirms the block of transactions and adds it to the blockchain.

10. The miner receives bitcoin rewards.

11. The process repeats.

Boom. Once I found something I could relate to blockchain, it was no longer intimidating. I felt confident that I could learn it and become a pro at it one day! It was hard work, but I learned to be patient with myself and not beat myself up when I couldn't grasp a concept right away. But the more time I studied it, the smarter I got on it—I felt the excitement again. Now, three years later, I teach blockchain.

I even have a website devoted to blockchain education where people can go and learn blockchain in an easy (and fun) way.

Don't Let Someone Else Live Your Dream

You also probably have a dream that seems unlikely or impossible. I won't be one of those guys offering clichéd one-liners like *chase your dreams*. Deep down, you already know that you should chase your dreams. But I will challenge you to view your dream from a different perspective. Ask yourself, in five, 10, 15, or even 20 years could you live with yourself knowing that you didn't step out in faith to write that book, start that business, or invent that product? Think about how tormenting the regret would be. Even worse, think about how it would feel to see someone else who *did* step out in faith live your dream. God put dreams in our hearts for a reason—they are yours, not someone else's. Don't let fear stop you from pursuing it. Take one step at a time.

Don't be afraid of what will happen if you fail—be afraid of what will happen if you don't try.

Now that I've scared you, go out and do great things!

CRYPTO RESEARCH TECHNIQUES

Y ou've all heard the stories (or it's happened to you), investors losing tons of dough on lame crypto projects, ICO and DeFi exit scams, and pump-n-dump schemes. The crypto community is chock full of them. Yet despite investors losing millions, there's still not a lot of guidance to help investors spot these disreputable projects before they put all their money into them.

I want to help solve this problem and show you how to spot bad investments immediately. I created this step-by-step guide to teach you how to conduct cryptocurrency research. My goal here is to give you the tools to protect yourself from scammers and lame projects so that these unfortunate events won't happen to you.

PROJECT ADMINISTRATIVE – FOUNDATION

Understanding a crypto project's foundation is probably cryptocurrency research's most important (and time-consuming) aspect. This is where you need to strap in and pay close attention to what makes a project purposeful and sustainable. This goes for the project's purpose, road map, partners, website, competition, social outreach, and more.

Why this Is Important

How well a project's foundation performs will dictate the overall legitimacy of the project. Think about a project's foundation as the rudder that steers the entire project. If the project's white paper is poorly written, purpose and goals unclear, and the webpage is undeveloped, awkward and suspicious; you could be dealing with an abandoned project or a scam. Conversely, if the project's white paper is well written, its purpose clear, and website clean and updated constantly, then that is a good sign that the project has some legitimacy.

What You Need to Look For

The White Paper: Think of the white paper as a project's operating manual. The white paper will outline the project's products, visions, goals, technologies, blockchain structure, consensus mechanism, tokenomics, and so on. White papers can be excruciatingly technical, but you want to make sure that what the project team is proposing in the white paper makes sense.

The Product and Purpose: "But, Do We Need That?" This is what you need to ask yourself when reading about the problem the project is claiming to solve. A project's purpose is the reason why the project was created in the first place. When analyzing a project's purpose, you want to look for two things:

1. The problem that the project is claiming to solve

2. The purpose's overall clarity

Look at this critically because you may run into projects that claim to solve a problem that doesn't exist or a project whose purpose is not very clear.

Let's take Bitcoin, for example. Bitcoin was created to solve an apparent problem, the double spending problem that affected consumers around the world. Easy, right? So, again, the next time you come across an interesting project, ask yourself: "Do we need that?" The answer could be the difference between a legit project and a scam.

Strategy: *Ask yourself the same questions that project developers might ask themselves before developing a project.*

The Goals: When analyzing the project's goals, see if you can find the answer to these questions:

- Are the project's goals clear or are they chock full of holes?

- How long will it take for them to reach those goals?

- Are they obtainable?

- Are they too ambitious?

- Not ambitious enough?

If you can easily find satisfying answers to these questions, then that's a plus. If not, then tread carefully.

Location: Where does the project claim to be headquartered?

It's typically okay for a project's headquarters to be listed in a country that you do not live in, but you still want to pay attention to a few things:

1. Make sure there is an address listed for the project's headquarters.

2. Ensure the address is not listed in a country where cryptocurrency is banned.

The Website: Is there a project website?

Having a website is crazy vital because it is likely how the project team notifies the world that their project exists. When analyzing the project website, you want to make sure that:

1. It looks professional.

2. It's generally free of grammar/spelling errors.

3. Every webpage is complete (no unfinished pages).

4. Most importantly, ensure you can easily find answers to obvious questions like the project's description, the problem they claim to solve, the white paper, address of their headquarters, and members of their project team.

Note: A shoddy website can be a red flag for a scam or lame project so be aware!

Partnerships: Who is the project claiming to partner with? Do you think this is a legit partnership?

Project developers are notorious for announcing partnerships with a party they do not have a *confirmed* partnership with. They may have only met with an interested party or tailored their social media post to make it seem like they are in partnership with another party.

One strategy you should use is to confirm the partnership on the alleged partner's website or social media page. If you can't find anything, then reach out to them on social media and ask for a link to the news report, email, or something that proves the partnership exists.

Social Media Outreach: Are they on social media?

I don't think I need to explain how crucial social media is for a project's business. Any project can set up a social media account, but what you want to look for is the project's social media *presence*:

1. How active are they on social media?

2. How many followers do they have?

3. Are they interacting with their followers?

Check how often the project posts on its leading social media platform, whether it be Twitter, Reddit, Telegram, or whatever. A page with fewer than three posts a week might be a red flag so be aware.

Competing Projects: How does this project differ from a competitor's project? Is the project just a carbon copy of an existing project?

There will be times when you come across projects that offer the same type of service as another project but go about it differently. For example, blockchain oracle projects *Chainlink* and *Tellor* are both blockchain oracle platforms, but they work at two different speeds and implement two different mechanisms to validate data.

Carbon copy projects are smaller projects that mimic (almost precisely) larger, more established projects. They may even use them in their project name. For example, Shiba Inu is the most popular cryptocurrency meme coin. However, since Shiba Inu exploded in popularity and price, you can't be on Coin Market Cap for 30 seconds without seeing that dozens of other projects use the words "Shiba" or "Inu" their title (Bitshiba, Shiba Fantom, Hachiku Inu, Solana Inu). There is nothing wrong with two or more projects sharing the same goals, but there has to be *something* different and worthwhile between the two.

PROJECT TEAM

Now, we move from the foundational administrative side of the project to the people behind the project—the project team. Here, you simply want to dive into the backgrounds of each team member and verify their credentials to the best of your ability. This does NOT need to be a labor-intensive process, but you want to gather data on members' education, experience and work history, and the team's working dynamics.

Why this Is Important

A crypto project is only as good as its team. Once you've been in this space for a while, you will see many projects fail because the team fails. And unfortunately, people have a way of embellishing the truth about themselves and their credentials, so you must find out if the people behind the project you want to invest in are as legit as they say they are.

Where To Start

You should be able to find a list of project team members somewhere on their webpage with links to their personal Linked In profiles or Facebook pages. This, however, will vary among projects.

Note: Some crypto project members remain anonymous. This does not necessarily mean that the project is a scam, but you probably want to look at the project a bit harder before investing.

What To Look For

Verify Their Credentials – You want to ensure the team's education and experience match their position titles. For example, you should expect the person in charge of the project's technical development to have a degree in computer science from an accredited university and previous technical experience.

💡 *Note: It is not uncommon for members on smaller projects to take on several roles. This is fine as long as they are qualified to perform their duties.*

Team Dynamics – Each team member should bring a necessary skill to the project. For example, a good team might have an overall team manager that directs the movement of the project, a person(s) in charge of technical development and operations, a person(s) in charge of social media or marketing, a person(s) in charge of finance, and a person(s) in charge of the web development (if there is one). Again, small teams may have a person covering two or three of these roles, which is fine, but be sure that these project components are covered.

THE TECHNOLOGY

The technology is the heart of a crypto project, but unfortunately, this is where many investors tap out. Trying to understand how blockchain, smart contracts, and the various Dapps built on top work can send your brain into overdrive. However, if you want to know how a particular project operates, you must understand its technology; there is no way around this. The technology allows the project to function in the fashion it was created to perform. Without the relevant technological components in place, a project will fail.

What To Look For

Consensus Mechanism: This applies to projects that have their own blockchain. Most projects do not fall into this category because they are built on an existing blockchain like Ethereum or Solana. However, for projects with their own blockchain, its consensus mechanism is one of the most important components because it can determine how fast (or slow) the blockchain operates and how vulnerable it is to hacks.

Consensus simply means "agreement." Blockchains are decentralized networks; therefore, participants on a blockchain must collectively agree that a transaction is valid before proceeding with confirmations and finalizations (among other things).

That being said, you want to ensure that the project's consensus mechanism complements the project's purpose. For example, a speedy consensus mechanism might be in order if the project's purpose is to process cryptocurrency payments. However, if the project's operation doesn't require a lot of speed but utility and network protection, then a consensus mechanism like proof-of-work might suffice.

Tokenomics: A project's tokenomics explains the various purposes and use cases of the project's token or coin. This includes how the token functions on a blockchain, overall token supply, token burning, governance, staking, mining, and vesting schedules. You will find a project's tokenomics in the white paper 99% of the time.

Tokenomics varies greatly depending on the project; analyzing each one will be overwhelming, if not impossible. However, there are some basic questions to ask yourself when analyzing a project's tokenomics:

- Are tokens/coins used to pay for transactions or another service?

- Can you stake the token/coins to earn interest?

- Is there a burn mechanism? If so, how, when, and under what circumstance will they be burned?

- Are the tokens/coins used to vote on proposals?

- How are tokens/coins distributed among participants?

- What is the project's tokens/coins vesting schedule?

Again, the project's white paper should have these answers. Once you've found answers to these questions, ask yourself, *"Do their tokenomics make sense?"* If so, that's good. If not, then proceed with caution.

MARKET DATA

Okay, onto the fun part, the market data. Market data is where you will find stats on a project's finances. I chose to cover market data last simply because too many investors base their investment decisions solely on a project's market cap and trading volume and neglect the other aspects of a project. Conducting project and technology research first might make investors think twice before making a bad investment based on market data alone.

💡 *Note: For your reference, a great place to find market data projects and tokens is Coin Market Cap. I find them to have the most accurate and real-time market information.*

Why This Is Important

Market data can tell you how well (or unwell) a crypto project performs and how they measure up against similar projects.

Note: If you're struggling to find the market cap of a project, you can calculate it by multiplying the number of circulating tokens/coins by the price of the token/coin.

What To Look For

Market Cap - The market cap measures the overall monetary value of a project.

For example, if crypto project XYZ has a market cap of $1,000,000, then the project is worth $1,000,000. Larger market cap projects ($300M and larger) are generally perceived as stable and legitimate. This is important because it allows the investor (you) to weed out low market cap project scams and dead projects.

Note: A project with a small market cap doesn't mean that the project is dead or a scam. Some projects struggle because they haven't gotten all together yet, and this is fine for the short-term —all projects start somewhere. Just be sure to evaluate the project as a whole and not just market data.

Trading Volume – Trading volume is the number of assets (in this case, tokens) traded between buyers and sellers on an exchange at any given time. It's important to carefully analyze a token's trading volume because it indicates whether people are interested in investing in it. Knowing this, when the trading volume of a token is high, you can reasonably assume that there is great interest in that token. When the trading volume is low, you can assume the opposite.

Note: There are caveats to this. Some projects artificially pump up their token's volume to attract genuine investors. This is the epitome of a scam or pump-and-dump scheme, so proceed cautiously.

Market Cap-Trading Volume Relationship – Analyzing the market cap and trading volume separately is fine as long as you know what you are doing. However, for research purposes, it's best to analyze these two data points as a pair to ensure that the trading volume justifies the market cap.

Investors sometimes use different standards to determine if a project's market cap and trading volume warrant an investment. However, as a general rule of thumb (which may change based on market conditions), if the trading volume is less than 15% of the project's total market cap, it might indicate little to no market interest in that project.

For example: If crypto project XYZ has a market cap of $1,000,000 and a 15-day trading volume of 50,000, then the trading volume is only 5% of the market cap. This indicates low market interest in XYZ, and investors hardly trade it.

Feel free to set your own criteria when determining a project's worthiness, but be cautious of low market cap-trading volume relationships.

Crypto Exchanges – The more exchanges a token is listed on, the better. This is because exchanges give exposure to a cryptocurrency. Think of an exchange as an advertisement platform for a token.

For example, if token XYZ is listed on 25 exchanges, then it will be advertised to 25 times as many people as it would be if it was listed on one exchange. Many (not all) scam tokens are listed on fewer than five exchanges—usually low-quality exchanges, DEXs, and crypto launchpads.

💡 ***Note:*** *A token or coin traded on a few exchanges does not necessarily disqualify the project. There are plenty of hidden gems out there; just be sure to look at the project as a whole and not just its market data.*

Final Word

Do not expect to become a pro at crypto research overnight. Scams are out there, and they are becoming increasingly cunning—it is up to you to know how to make informed decisions based on solid research (that you conduct on your own). It will take some time to nail down these research principles and get good at separating the gems from the fakes but remember this: The more work you put in early, the less disappointed you will be later.

BLOCKCHAIN AND CRYPTOCURRENCY GLOSSARY

Address – A string of letters and numbers that represents the place where cryptocurrency is stored [same as wallet address]

AML – Anti Money Laundering – Standards and procedures governments implement to prevent money from being laundered physically or electronically

AMM – Automated Market Maker. A smart contract that uses an algorithm to set the price of a cryptocurrency in lieu of an order book

API – Application Program Interface. Software that enables other software, computer programs, and applications to interact

Application Program Interface (API) – Software that enables other software, computer programs, and applications to interact

Application Specific Integrated Circuit (ASIC) – A computer chip specifically used to solve SHA-256 puzzles during mining

ASIC – Application Specific Integrated Circuit. A computer chip specifically used to solve SHA-256 puzzles during mining

Automated Market Maker (AMM) – A smart contract that uses an algorithm to set the price of a cryptocurrency in lieu of an order book

Base 58 – An alphanumeric numbering system where a digit in an address is represented by one of 58 letters and numbers excluding "I," "L," "O," and "U"

Bitcoin – A blockchain network that facilitates bitcoin and Bitcoin Lightning Network transactions

bitcoin – A digital currency used to pay for goods and services and used as an investment vehicle and store of value

Bitcoin Lightning Network – A smart contract-based bitcoin payment system that enables bitcoin micropayments between users

Block – a digital storage box where data is stored

Block Header – The block's "headline" that describes what's inside of a block, including the block version, hash value, hash value of the previous block, timestamp, difficulty level, nonce, and the merkle root

Block Height – Describes the highest numbered block on a blockchain

Bug – An exploitable opening in a smart contract's code

Central Processing Unit (CPU) – A computer (PC or Laptop)

Cold Wallet – A manual hardware device that you can physically store in the safety of your own home or another secure location

Consensus Mechanism – A process that nodes in a network follow to collectively agree on the validity of a transaction or block

CPU – Central Processing Unit. A computer (PC or Laptop)

Crowdfunding – A process new blockchain projects undergo to raise money from public sources, typically people and companies interested in the project

Crowdloan – An activity where individuals are rewarded (in cryptocurrency) for loaning their cryptocurrency to a project for a specified amount of time, after which their initial investment will be returned with the reward.

Crowdsourcing – A process blockchain projects undergo to gather data or money from public sources

Crypto Exchange – A digital marketplace through which users buy, sell, or swap cryptocurrency

Crypto Twitter – A network of cryptocurrency project developers, traders, and enthusiasts who post their thoughts on trending cryptocurrency topics on Twitter

Crypto Wallet – [Slang for hardware wallet.] A digital wallet where cryptocurrency is stored

Cryptocurrency – Digital money that runs on a blockchain and is used to pay for goods and services offered on and off the blockchain

Cryptocurrency Address – A string of letters and numbers that users use to send or receive cryptocurrency transactions

Cryptography – The process of securely sending messages to and from another party in a way that no unauthorized person can decipher

Custodial Wallet – A cryptocurrency wallet that is owned and operated by a central entity like a centralized cryptocurrency exchange

DAO – Decentralized Autonomous Organization. A body of computer codes created by a group of like-minded individuals to automatically execute tasks on a blockchain

Dapp – Decentralized Application. An open-source computer application that runs on a blockchain or another peer-to-peer network

Decentralized Application (Dapp) – An open-source computer application that runs on a blockchain or another peer-to-peer network

Decentralized Autonomous Organization (DAO) – A body of computer codes created by a group of like-minded individuals to automatically execute tasks on a blockchain

Decentralized Exchange (DEX) – A decentralized application through which users can swap, lend, and borrow cryptocurrency

Decentralized Finance, or DeFi – A blockchain-based peer-to-peer financial system that offers a wide range of conventional financial services that is typically found at a traditional bank, the stock market, or other financial institutions

Delegate Proof-of-Stake – A consensus mechanism through which users can delegate other participants to be validators on the blockchain

DEX – Decentralized Exchange. A decentralized application through which users can swap, lend, and borrow cryptocurrency

Difficulty – An estimate that describes how difficult it will be to mine a block

DPoS – Delegated Proof-of-Stake. A consensus mechanism through which users can delegate other participants to be validators on the blockchain

ECDSA – Elliptic Curve Digital Signature Algorithm. A type of digital signature algorithm used to sign transactions on a blockchain

EIP – Ethereum Improvement Proposal. Describe standards for the Ethereum platform, including core protocol specifications, client APIs, and contract standards

EIP-1559 – An Ethereum upgrade that increases block sizes and establishes a base fee for Ethereum transactions. Users will be able to set the price required to complete transactions on the Ethereum blockchain

Elliptic Curve Digital Signature Algorithm – A type of digital signature algorithm used to sign transactions on a blockchain

ERC – Ethereum Request for Comment. A type of token standard that describes how Ethereum tokens should be issued and operate on the Ethereum blockchain

ERC-20 – Derived from EIP-20, an Ethereum token standard enables tokens to be transferred and approved so another on-chain third party can spend them

ERC-721 – A Non-Fungible Token (NFT) standard. A standard interface that allows wallet/broker/auction applications to work with any NFT on Ethereum

Ethereum 2.0 – An Ethereum fork that will convert Ethereum's Proof-of-Work consensus mechanism to a Proof-of-Stake consensus mechanism and make Ether a deflationary asset

Ethereum Improvement Proposal (EIP) – Describes standards for the Ethereum platform, including core protocol specifications, client APIs, and contract standards

Ethereum Request for Comment (ERC) – A type of token standard that describes how Ethereum tokens should be issued and operate on the Ethereum blockchain

Exchange – A digital marketplace through which users can buy and sell assets

Field Programmable Gate Array (FPGA) – A computer chip that can be programmed to solve cryptographic puzzles during mining

FOMO – Fear of Missing Out

FPGA – Field Programmable Gate Array. A computer chip that can be programmed to solve cryptographic puzzles during mining

Genesis Block – The very first block in a blockchain

Graphical Processing Unit (GPU) – A processing unit designed for graphics and video rendering but used in blockchain mining rigs

GPU – Graphical Processing Unit. A processing unit designed for graphics and video rendering but used in blockchain mining rigs

Hash – A hash is a fixed-length string of letters and numbers that represents a piece of data

Hash Rate – As bitcoin mining increases, the hash rate increases because the difficulty increases with more users on the network

Height – Block Height. Describes the highest block number on a blockchain

Hot wallet – [Slang for *software wallet.*] A software-based wallet that comes in the form of a downloadable mobile or desktop app

ICO – Initial Coin Offering. Occurs when cryptocurrency project developers sell their first lot of cryptocurrency to the public to raise funds

IDO – Initial Dex Offering. Occurs when cryptocurrency project developers list their first lot of cryptocurrencies on an existing decentralized cryptocurrency exchange (DEX) to raise funds for their project

IEO – Initial Exchange Offering. Occurs when cryptocurrency project developers list their first lot of cryptocurrencies on an existing crypto exchange to raise funds for their project

Initial Coin Offering (ICO) – Occurs when cryptocurrency project developers sell their first lot of cryptocurrencies directly to the public to raise funds for their project

Initial Dex Offering (IDO) – Occurs when cryptocurrency project developers list their first lot of cryptocurrencies on an existing decentralized cryptocurrency exchange (DEX) to raise funds for their project

Initial Exchange Offering (IEO) – Occurs when cryptocurrency project developers list their first lot of cryptocurrencies on an existing crypto exchange to raise funds for their project.

Inputs – The amounts of bitcoin sent from one wallet to another

Internet Of Things (IoT) – Describes the interconnectivity of physical devices like smartphones and smartwatches to other devices like small appliances, smart doors, and smart TV through the internet or web-based applications.

IoT – Internet Of Things. Describes the interconnectivity of physical devices like smartphones and smartwatches to other devices like small appliances, smart doors, and smart TV through the internet or web-based applications

Know Your Customer (KYC) – A standard procedure financial institutions must follow to verify the identities of clients with whom they intend to do business or serve

KYC – Know Your Customer. A standard procedure financial institutions must follow to verify the identities of clients with whom they intend to do business or serve

layer 1 – [Slang for *blockchain.*] The bottom layer (the blockchain) of a blockchain application

layer 2 – [Slang for *smart contract.*] A smart contract that runs on top of a blockchain

Leaf node – A data point on the merkle tree that contains a list of transaction hashes

Liquidity Pool – A DeFi protocol where users (liquidity providers) deposit cryptocurrency pairs to earn compound interest

Liquidity Provider – A DeFi user that deposits cryptocurrency pairs into a liquidity pool

London Hard Fork – An Ethereum upgrade that will change Ethereum's Proof-of-Work consensus mechanism to a Proof-of-Stake consensus mechanism and make Ether a deflationary asset

Market Cap – The market cap measures the overall monetary value of a project

Memory (mem) pool – An area on the blockchain where unconfirmed transactions are stored until they are added to a block

Merkle Root – a data point in the merkle tree that contains every leaf node, parent node, and transaction inside of a block

Merkle Tree – a binary hierarchal data structure that contains an index of transactions stored inside of a block

Miner – A computer node on a blockchain that solves cryptographic puzzles for the chance to earn cryptocurrency rewards

Mining – The process of solving a cryptographic puzzle for the chance to earn cryptocurrency rewards

Mining Farm – A collection of mining rigs based in one location to mine cryptocurrency

Mining Pool – A body of nodes that combine computation power to increase their chance of successfully solving a cryptographic puzzle and earning cryptocurrency rewards

Mining Rig – Hardware built specifically for mining cryptocurrency

Mining Software – Downloadable software that governs how mining rigs operate during mining

NFT – Non-Fungible Token. A unique digital asset made to represent any real-world item and that cannot be counterfeited, reproduced, or divided

Node – A computer connection point in a blockchain network that records and transmits data, validates transactions, adds blocks to the chain, and secures the network

Non-Custodial Wallet – A decentralized cryptocurrency wallet whose assets are wholly owned by the wallet user

Non-Fungible Token (NFT) – A unique digital asset made to represent any real-world item and that cannot be counterfeited, reproduced, or divided

Oracle – A decentralized application used to pull in real-world data onto the blockchain

Outputs – The amount of bitcoin stored in a user's bitcoin wallet

Paper Wallet – A piece of paper that contains a cryptocurrency address and private key to store cryptocurrency

Permissioned – A system that grants selected users rights and privileges to the blockchain

Permissionless – A system that grants everyone rights and permissions to use the blockchain

Play-To-Earn – A method by which gamers are paid cryptocurrency to play blockchain-based videogames to collect rare or valuable in-game items (NFTs)

PoA – Proof of Authority. A consensus mechanism where validators are required to stake their reputations for the opportunity to validate blocks and earn cryptocurrency rewards

PoAC – Proof-of-Activity. A consensus mechanism that combines elements of the proof-of-work consensus mechanism with aspects of the proof-of-stake consensus mechanism

PoC – Proof-of-Capacity. A consensus mechanism where nodes on a network utilize the free space on their hard drives to store pre-calculate responses to proof-of-work puzzles

PoS – Proof-of-Stake. A blockchain consensus mechanism by which node participants deposit, or stake, tokens into a node to earn the opportunity to confirm (forge) a block and receive rewards

PoW – Proof-of-Work. A blockchain consensus mechanism by which miners in the network compete by solving a cryptographic puzzle for the opportunity to confirm a block and receive block rewards

Prediction Market – A decentralized marketplace where users can find a catalog of current world events and place bets on the likely outcome of those events

Private Blockchain – A blockchain explicitly developed for private use, typically in a centralized organization

Private key – A single-key encryption mechanism by which one key is used to both encrypt and decrypt data

Proof of Authority (PoA) – A consensus mechanism where validators are required to stake their reputations for the opportunity to validate blocks and earn cryptocurrency rewards

Proof-of-Activity – (PoAC) A consensus mechanism that combines elements of the proof-of-work consensus mechanism with aspects of the proof-of-stake consensus mechanism

Proof-of-Capacity (PoC) – A consensus mechanism where nodes on a network utilize the free space on their hard drives to store pre-calculate responses to proof-of-work puzzles

Proof-of-Stake (PoS) – A blockchain consensus mechanism by which node participants deposit, or stake, tokens into a node to earn the opportunity to confirm (forge) a block and receive rewards

Proof-of-work (PoW) – A blockchain consensus mechanism by which miners in the network compete by solving a cryptographic puzzle for the opportunity to confirm a block and receive block rewards

Public Blockchain – A blockchain that is designed for public use and whose data is available for the public to view

Public Key – A two-key encryption mechanism by which one key is used to encrypt data, and a separate key is used to decrypt data on a blockchain

Smart Contract – A body of computer codes programmed to execute tasks on a blockchain automatically

Software Wallet – A software-based wallet that comes in the form of a downloadable mobile or desktop app

Token – A type of cryptocurrency that does not have its own blockchain

Tokenomics – Explains the various purposes, functions, and use cases of a project's coin or token

Trading Volume –The number of assets (in this case, tokens) being traded between buyers and sellers on an exchange at any given time

Trustless – Negates the need to place complete trust into one blockchain node

Trustless Verification – A concept used in blockchain which negates a user's need to rely on a single blockchain node to validate a transaction

Turing-Complete – A computer program that can run many types of other computer programs using loops and recursion

Turing-Incomplete – A computer program in which all aspects and outcomes of its processes are predictable

Unspent Transaction Outputs (UTXO) – The amount of bitcoin stored in a user's bitcoin wallet

UTXO – Unspent Transaction Outputs. The amount of bitcoin stored in a user's bitcoin wallet

Validate – A process by which nodes on a blockchain network collectively attest to the validity of a transaction

Validation Process – A process by which nodes on a blockchain network collectively attest to the validity of a transaction

Validator – A person, typically a node or a mine, that validates transactions on a blockchain

Verify – The process of confirming that an action or process has taken place

Wallet – A software or hardware location where cryptocurrency is stored

Wallet Address – A string of letters and numbers that users use to send or receive cryptocurrency transactions

Web 1.0 – A term that describes the inception of the internet

Web 2.0 – A term that describes the era of using the internet to connect people through social media and applications

Web 3.0 – A term that describes the era of decentralized internet and peer-to-peer connectivity

Whitepaper – A document that details a blockchain project's products, vision, goals, technology stacks, consensus mechanism, and tokenomics

Zero Knowledge Proof – A method used to prove that a piece of information is authentic without revealing the actual information

ENDNOTES

1   Satoshi Nakamoto, *Bitcoin: A Peer-to-Peer Electronic Cash System*, 2008, https://bitcoin.org/bitcoin.pdf.

2   Paulina Likos and Coryanne Hicks, "The History of Bitcoin, the First Cryptocurrency," U.S. News & World Report, February 4, 2022, https://money.usnews.com/investing/articles/the-history-of-bitcoin.

3   Pete Rizzo, "10 Years Ago Today, Bitcoin Creator Satoshi Nakamoto Sent His Final Message," Forbes, April 26, 2021, https://www.forbes.com/sites/peterizzo/2021/04/26/10-years-ago-today-bitcoin-creator-satoshi-nakamoto-sent-his-final-message/.

4   Patrick J. Connolly, "John Locke," n.d., https://iep.utm.edu/locke.

5   Jamie Bartlett, "Cypherpunks Write Code," American Scientist, February 6, 2017, https://www.americanscientist.org/article/cypherpunks-write-code.

6   "Timothy C. May—Thirty Years of Crypto Anarchy," YouTube video, February 17, 2017, https://www.youtube.com/watch?v=TdmpAy1hI8g.

7   Arvind, Narayanan, "What Happened to the Crypto Dream? Part 1." *IEEE Security & Privacy* 11, no. 2 (March 2013): 75–76.

8   David L. Chaum, "Security without Identification: Transaction Systems to Make Big Brother Obsolete." *Communications of the ACM* 28, no. 10, (October 1, 1985): 1030–1044. https://doi.org/10.1145/4372.4373.

9   David L. Chaum, "Untraceable Electronic Mail, Return Addresses, and Digital Pseudonyms." *Communications of the ACM* 24, no. 2 (February 1, 1981): 84–90.

10   Video: David Chaum, Creator of xx, Shares his Experiences Tangling with the NSA, YouTube, Aug 3, 2020, https://www.youtube.com/watch?v=5ig-3G5jEXQ, Marked at 5:30.

11   David Chaum, "Untraceable Electronic Mail, Return Addresses, and Digital Pseudonyms," *Communications of the ACM* 24, no. 2 (1981): 84–90, https://doi.org/10.1145/358549.358563.

12   David Chaum, "Blind Signatures for Untraceable Payments," in *Advances in Cryptology Proceedings of Crypto 82*, eds. Ronald L. Rivest, Alan T. Sherman and David Chaum (New York: Springer-Verlag, 1983), 199-203, http://www.chaum.com/publications/publications.html

13   David L. Chaum, Amos Fiat, Moni Naor, "Untraceable Electronic Cash." In *Advances in Cryptology — CRYPTO' 88*, ed. Shafi Goldwasser, (New York, NY: Springer, 1990), 319–27.

14   Chaum, David. "Blind Signatures for Untraceable Payments." In *Advances in Cryptology*, ed. David Chaum, Ronald L. Rivest, and Alan T. Sherman, (Boston, MA: Springer, 1983): 199–203.

15   Jay Stowsky, "Secrets or Shields to Share? New Dilemmas for Dual Use Technology Development and the Quest for Military and Commercial Advantage in the Digital Age." UCAIS Berkeley Roundtable on the International Economy, Working Paper Series. UCAIS Berkeley Roundtable on the International Economy, UC Berkeley, February 21, 2003. https://econpapers.repec.org/paper/cdlucbrie/qt89r4j908.htm.

16   Title 22. U.S.C. CFR § 121. July 22, 1993; Kenneth J. Pierce (1984) "Public Cryptography, Arms Export Controls, and the First Amendment: A Need for Legislation," *Cornell International Law Journal*: Vol. 17 : No. 1 , Article 5.

17   Kenneth Pierce, "Public Cryptography, Arms Export Controls, and the First Amendment: A Need for Legislation." *Cornell International Law Journal* 17, no. 1 (January 1, 1984): 197–236.

18   "Unchained Podcast, Why Bitcoin Now: David Chaum and Adam Back Reflect on the Crypto Wars - Ep.186," YouTube video, August 18, 2020, https://www.youtube.com/watch?v=ZVZxRMAeIdo.

19   Lee Ann Gilbert, "Patent Secrecy Orders: The Unconstitutionality of Interference in Civilian Cryptography under Present Procedures." *Santa Clara Law Review* 22, no. 2 (January 1, 1982): 325.

20   Ibid.

21 R. L. Rivest, A. Shamir, and L. Adelman, "A Method for Obtaining Digital Signatures and Public-Key Cryptosystems." *Communications of the ACM* 21, no. 2 (February 1978): 120–26. https://doi.org/10.1145/359340.359342.

22 Kevin McCurley et al., "History of the IACR"; International Association for Cryptologic Research, 2022, https://iacr.org/docs/history.

23 "David Chaum shares the history of Digital currency – Deconomy," Accessed August 11, 2022. https://deconomy.com/tokyo-institute-of-technology-taps-for-fastest-ai-supercomputer.

24 Alan Gersho., "Paper: Advances in Cryptology: A Report on CRYPTO 81." Accessed August 11, 2022. https://www.iacr.org/cryptodb/data/paper.php?pubkey=23746.

25 "Unchained Podcast, Why Bitcoin Now: David Chaum and Adam Back Reflect on the Crypto Wars - Ep.186," YouTube video, August 18, 2020, https://www.youtube.com/watch?v=ZVZxRMAeIdo.

26 Kevin McCurley et al., "History of the IACR"; International Association for Cryptologic Research, 2022, https://iacr.org/docs/history.

27 David L. Chaum, "ECASH – Chaum.Com." Accessed April 8, 2022. https://chaum.com/ecash.

28 Timothy C. May, *The Cryptoanarchist Manifesto*, 1988

29 Anton P., "Cypherpunk ideas, principles, and influence on digital society," June 18, 2021, https://atlasvpn.com/blog/cypherpunk-ideas-principles-and-influence-on-digital-society.

30 Sameer Parekh, "Prospects for Remailers," First Monday, August 5, 1996, https://doi.org/10.5210/fm.v1i2.476.

31 Timothy May, "The Cyphernomicon: Cypherpunks FAQ and More, Version 0.666," n.d., https://nakamotoinstitute.org/static/docs/cyphernomicon.txt.

32 Eric Hughes., "A Cypherpunk's Manifesto," March 9, 1993, https://www.activism.net/cypherpunk/manifesto.html.

33 What is the Cypherpunks Mailing List?, cryptoanarchy.wiki, February 20, 2022. https://cryptoanarchy.wiki/getting-started/what-is-the-cypherpunks-mailing-list. (WIKIPEDIA ARTICLE)

34 Ibid.

35 Anton P., "Cypherpunk ideas, principles, and influence on digital society," June 18, 2021, https://atlasvpn.com/blog/cypherpunk-ideas-principles-and-influence-on-digital-society.

36 Adam Back, "Hashcash—A Denial of Service Counter-Measure," August 01, 2002, http://www.hashcash.org/hashcash.pdf.

37 Satoshi Nakamoto, "Bitcoin: A Peer-to-Peer Electronic Cash System," 2008, https://bitcoin.org/bitcoin.pdf.

38 Aaron van Wirdum, "The Genesis Files: If Bitcoin Had a First Draft, Wei Dai's B-Money Was It," *Bitcoin Magazine*, June 14, 2018, https://bitcoinmagazine.com/technical/genesis-files-if-bitcoin-had-first-draft-wei-dais-b-money-was-it.

39 Ibid.

40 Nick Szabo, "Bit Gold," *Satoshi Nakamoto Institute*, December 29, 2005, https://nakamotoinstitute.org/bit-gold.

41 Satoshi, post to "Re: They want to delete the Wikipedia article," Satoshi Nakamoto Bitcoin Forum July 20, 2010, 6:38 p.m., https://bitcointalk.org/index.php?topic=342.msg4508#msg4508.

42 Aaron van Wirdum, "The Genesis Files: How Hal Finney's Quest for Digital Cash Led to RPOW (and More)," *Bitcoin Magazine*, August 28, 2020, https://bitcoinmagazine.com/culture/the-genesis-files-how-hal-finneys-quest-for-digital-cash-led-to-rpow-and-more.

43 Hal Finney, "RPOW - Reusable Proofs of Work." Accessed August 11, 2022. https://cryptome.org/rpow.htm.

44 Andrea Peterson, "Hal Finney Received the First Bitcoin Transaction. Here's How He Describes It," *Washington Post*, January 3, 2014, https://www.washingtonpost.com/news/the-switch/wp/2014/01/03/hal-finney-received-the-first-bitcoin-transaction-heres-how-he-describes-it/.

45 "Timothy C. May—Thirty Years of Crypto Anarchy | HCPP16," YouTube video, February 11, 2017, https://www.youtube.com/watch?v=TdmpAy1hI8g.

46 "The Mailing List where Bitcoin Began, with Perry Metzger," YouTube video, September 10, 2018, https://www.youtube.com/watch?v=l0WXFhk3dnU.

47 Michael Hudson, "Now Greenspan Wants to Take It All Back," CounterPunch, March 31, 2011, https://www.counterpunch.org/2011/03/31/now-greenspan-wants-to-take-it-all-back/.

48 Chris Isidore, "Home Prices: 1st Drop in 11 years," CNN, September 25, 2006, https://money.cnn.com/2006/09/25/news/economy/homesales2.

49 "The U.S. Financial Crisis," Council on Foreign Relations, 2022, https://www.cfr.org/timeline/us-financial-crisis/.

50 Darryl, E. Getter, *Fannie Mae and Freddie Mac in Conservatorship: Frequently Asked Questions,* (R44525), Congressional Research Service, July 22, 2020, https://crsreports.congress.gov/product/details?prodcode=R44525.

51 Deborah Lucas, *Measuring the Cost of Bailouts*, November, 2018.

52   Kimberly Amadeo, "2008 Financial Crisis: Causes, Costs, and Whether It Could Happen Again," The Balance, February 10, 2022, https://www.thebalance.com/2008-financial-crisis-3305679.

53   Ibid.

54   "Cattle in the ancient world of the Bible," Women In The Bible, n.d., https://womeninthebible.net/bible_daily_life/cattle_ancient_world.

55   "Les Cauris," Citeco, May 27, 2013, https://www.citeco.fr/les-cauris.

56   Kallie Szczepanski, "The Invention of Paper Money in China," ThoughtCo, October 17, 2019, https://www.thoughtco.com/the-invention-of-paper-money-195167.

57   " History and Purpose," European Union, n.d., https://european-union.europa.eu/institutions-law-budget/euro/history-and-purpose_en.

58   "Countries Using the Euro," European Union Website, n.d., https://european-union.europa.eu/institutions-law-budget/euro/countries-using-euro_en.

59   "What is Intrinsic Value? Definition and Examples," Market Business News, n.d., https://marketbusinessnews.com/financial-glossary/intrinsic-value-definition-meaning/#:~:text=For%20put%20options%20it%20is,option's%20intrinsic%20value%20is%20%2411.

60   Allen R. McConnell, Robert J. Rydell, Laura M. Strain, Diane M. Mackie, Forming Implicit and Explicit Attitudes Toward Individuals: Social Group Association Cues, *Journal of Personality and Social Psychology* 94, no. 5 (May 2008): 792–807. https://doi.org/10.1037/0022-3514.94.5.792.

61   Nalini Ambady and Robert Rosenthal, "Half a Minute: Predicting Teacher Evaluations from Thin Slices of Nonverbal Behavior and Physical Attractiveness." *Journal of Personality and Social Psychology* 64, no. 3 (1993): 431–41. https://doi.org/10.1037/0022-3514.64.3.431.

62   Gul Gunaydin, Emre Selcuk, and Vivian Zayas, "Impressions Based on a Portrait Predict, 1-Month Later, Impressions Following a Live Interaction." *Social Psychological and Personality Science* 8, no. 1 (January 2017): 36–44. https://doi.org/10.1177/1948550616662123.

63   "Confirmation Bias," American Psychological Association, n.d., https://dictionary.apa.org/confirmation-bias.

64   Yun Li, Y., "Fidelity Is Offering 401(k) Investors Access to Bitcoin, the First Retirement Plan Provider to Do So," CNBC, April 26, 2022, https://www.cnbc.com/2022/04/26/fidelity-offers-401k-investors-access-to-bitcoin-a-retirement-plan-first.html.

65   Jennifer Surane, "J.,Morgan Stanley Taps Sheena Shah to Lead New Crypto Research Team," Bloomberg, September 13, 2021, https://www.bloomberg.

com/news/articles/2021-09-13/morgan-stanley-taps-sheena-shah-to-lead-new-crypto-research-team.

66 Ibid.

67 Ibid.

68 James K. Jackson, ed. "Global Economic Effects of COVID-19: Overview - Congress." crsreports.congress.gov, September 28, 2021. https://crsreports.congress.gov/product/pdf/R/R46270.

69 Mieszko Mazur, Man Dang, and Miguel Vega. "COVID-19 and the March 2020 Stock Market Crash. Evidence from S&P1500." *Finance Research Letters* 38 (January 2021): 101690. https://doi.org/10.1016/j.frl.2020.101690.

70 James K. Jackson, *Global Economic Effects of COVID-19: Overview* (R46270), Congressional Research Service, February 14, 2022, https://crsreports.congress.gov/product/details?prodcode=R46270.

71 Bitcoin USD (BTC-USD), Yahoo! Finance, January 01, 2020–December 30, 2020.

72 Lennart Ante, "How Elon Musk's Twitter Activity Moves Cryptocurrency Markets," *SSRN Electronic Journal*, 2021. https://doi.org/10.2139/ssrn.3778844.

73 Steve Kovach, "Tesla Buys $1.5 Billion in Bitcoin, Plans to Accept It as Payment," CNBC, February 8, 2021, https://www.cnbc.com/2021/02/08/tesla-buys-1point5-billion-in-bitcoin.html.

74 "Tesla Q1 2021 Update, Statement of Cashflows (Unaudited)," Tesla, 26.

75 Tweet by @elonmusk, May 12, 2021, https://twitter.com/elonmusk/status/1392602041025843203.

76 Rishi Iyengar, "Bitcoin Plunges 12% After Elon Musk Tweets That Tesla Will Not Accept It as Payment," CNN Business, May 13, 2021, https://www.cnn.com/2021/05/12/tech/elon-musk-tesla-bitcoin/index.html.

77 Kevin Helms, "Elon Musk Confirms He Owns Bitcoin, Has Not Sold Any—Tesla Intends to Hold BTC Long Term, Sold Some to Prove Liquidity," Bitcoin.com, April 27, 2021, https://news.bitcoin.com/elon-musk-owns-bitcoin-has-not-sold-any-tesla-hold-btc-long-term-sold-to-prove-liquidity.

78 Bitcoin Mining Council. "Welcome to the Bitcoin Mining Council," n.d., http://bitcoinminingcouncil.com.

79 Tweet by @ elonmusk, June 3, 2021, https://twitter.com/elonmusk/status/1400620080090730501.

80   Omkar Godbole, "Bitcoin Drops After Musk Tweets of Breakup; Musk's broken heart sent BTC down nearly 7%", CoinDesk, June 4, 2021, Updated September 14, 2021, https://www.coindesk.com/markets/2021/06/04/bitcoin-drops-after-musk-tweets-of-breakup/

81   Tweet by @elonmusk, June 3, 2021, https://twitter.com/elonmusk/status/1400645833150840835.

82   Tweet by @elonmusk, April 2, 2019, https://twitter.com/elonmusk/status/1113009339743100929.

83   Tweet by @elonmusk, April 28, 2021, https://twitter.com/elonmusk/status/1387290679794089986.

84   Tweet by @elonmusk, July 17, 2020, https://twitter.com/elonmusk/status/1284291528328790016.

85   David Rodeck, "An Introduction to Dogecoin, The Meme Cryptocurrency," Forbes, April 20, 2021, https://www.forbes.com/advisor/investing/cryptocurrency/what-is-dogecoin/.

86   Omkar Godbole, "Dogecoin Eclipses XRP as 4th Largest Cryptocurrency Ahead of 'Dogeday'," Coindesk, April 19, 2021, https://www.coindesk.com/markets/2021/04/19/dogecoin-eclipses-xrp-as-4th-largest-cryptocurrency-ahead-of-dogeday.

87   Tweet by @elonmusk , May 13, 2021, https://twitter.com/elonmusk/status/1392974251011895300?s=20.

88   Erin Gobler, "What Is a Meme Coin?", The Balance, June 29, 2022, https://www.thebalance.com/what-is-a-meme-coin-5224632.

89   "Biz Cas Fri 1" YouTube video, September 29, 2009, https://www.youtube.com/watch?v=tLSgRzCAtXA.

90   Amanda Lenhart, Aaron W. Smith, Monica Anderson, Maeve Duggan, and Andrew Perrin. "Teens, Technology and Friendships." Pew Research Center, August 6, 2015. https://apo.org.au/node/56457.

91   Erin Gobler, "What Is a Bag Holder in Investing?", The Balance, June 29, 2022, https://www.thebalance.com/what-is-a-bag-holder-in-investing-5218858.

92   "Squid Game Crypto Token Collapses in Apparent Scam," BBC News, November 2, 2021, https://www.bbc.com/news/business-59129466.

93   Squid Game Transaction Hash: 0x0df9d3177ba8e642b6367b8 646597f0aa357ff677bca0c8a69e88318572ff53, BscSCan, November 1, 2021.

94   "Pump-and-Dump Scheme: What It Is and How to Avoid One," Bankrate, June 30, 2022, https://www.bankrate.com/investing/pump-and-dump-scheme/.

95 Brian Quarmby, "Someone Bought $3,400 Worth of SHIB Last August. It's Now Worth $1.55 Billion," Cointelegraph, October 28, 2021, https://cointelegraph.com/news/someone-bought-3-400-worth-of-shib-last-august-it-s-now-worth-1-55-billion.

96 Daniel Phillips, "Why Is Bitcoin's Supply Limit Set to 21 Million?", Decrypt, December 30, 2020, https://decrypt.co/34876/why-is-bitcoins-supply-limit-set-to-21-million.

97 Nicole Willing, "How Many Shiba Inu Coins Are There?", Capital.com, July 30, 2022, https://capital.com/how-many-shiba-inu-coins-are-there-in-the-world.

98 Dabiel Kahneman and Amos Tversky, "Prospect Theory: An Analysis of Decision Under Risk," *World Scientific Handbook in Financial Economics Series* (2013): 99–127.

99 Gregory Ciotti, "Color Psychology in Marketing and Branding Is All About Context," Help Scout, August 12, 2020, https://www.helpscout.com/blog/psychology-of-color/.

100 Kat Tretina, "10 Best Cryptocurrencies of August 2022," Forbes, August 1, 2022, https://www.forbes.com/advisor/investing/cryptocurrency/top-10-cryptocurrencies/.

101 Dashboard, Defi Lama, https://defillama.com.

102 Mike Antolin, "What Are Liquidity Pools?", CoinDesk, June 7, 2022, https://www.coindesk.com/learn/what-are-liquidity-pools/.

103 "Ethereum's New 1MB Blocksize Limit," BitMEX blog, December 2, 2021, https://blog.bitmex.com/ethereums-new-1mb-blocksize-limit/.

104 Raynor de Best, Ethereum (ETH) gas price history up until May 16, 2022, Statista, May 17, 2022, https://www.statista.com/statistics/1221821/gas-price-ethereum/

105 "Binance Smart Chain: Q3 2021 Overview," DappRader, October 7, 2021, https://dappradar.com/blog/binance-smart-chain-q3-2021-overview.

106 "Upgrading Ethereum to Radical New Heights," Ethereum.org, August 9, 2022, https://ethereum.org/en/upgrades.

107 "Gas and Fees," Ethereum.org, August 9, 2022, https://ethereum.org/en/developers/docs/gas/#eip-1559.

108 "Securities and Exchange Commission v. Trendon T. Shavers and Bitcoin Savings and Trust,"

(Litigation Release No. LR-23090), September 22, 2014 https://www.sec.gov/litigation/litreleases/2014/lr23090.htm.

109 The Securities and Exchange Commission, *SEC v. BTC Trading, Corp. and Ethan Burnside*, File No. 3-16307 (2014), https://www.sec.gov/litigation/admin/2014/33-9685.pdf.

110 The Securities and Exchange Commission, *SEC v. Erik T. Voorhees*, File No. 3-15902 (2014), https://www.sec.gov/litigation/admin/2014/33-9592.pdf

111 Simona Mola, *SEC Cryptocurrency Enforcement: Q3 2013–Q4 2020*, Cornerstone Research, 2021, https://www.cornerstone.com/insights/reports/sec-cryptocurrency-enforcement-q3-2021-update/.

112 Simona Mola, *SEC Cryptocurrency Enforcement: Q3 2013–Q4 2020*, Cornerstone Research, 2021, https://www.cornerstone.com/insights/reports/sec-cryptocurrency-enforcement-q3-2021-update/.

113 U.S. Securities and Exchange Commission, *Securities Act of 1933*, Legal Information Institute, 1933, https://www.law.cornell.edu/wex/securities_act_of_1933.

114 U.S. Securities and Exchange Commission, *Framework for "Investment Contract" Analysis of Digital Assets*, SEC.gov, https://www.sec.gov/files/dlt-framework.pdf.

115 *SEC v. Zachary Corbin*, F.3-18888 (S.D. of New York 2018).

116 *SEC v. Telegram Group Inc. and Ton Issuer Inc* 19 Civ. 9439 (PKC) (S.D. of New York 2019).

117 *SEC v. Ripple Labs, Inc., Bradley Garlinghouse, and Christian A. Larsen*, C 1:20-cv-10832 (S.D. of New York 2020).

118 *SEC Cryptocurrency Enforcement*, Cornerstone Research, 2021, https://www.cornerstone.com/wp-content/uploads/2022/01/SEC-Cryptocurrency-Enforcement-2021-Update.pdf.

119 U.S. Securities and Exchange Commission, "SEC Division of Examinations Announces 2022 Examination Priorities," SEC press release, 2022, https://www.sec.gov/news/press-release/2022-57.

120 U.S. Securities and Exchange Commission, "SEC Nearly Doubles Size of Enforcement's Crypto Assets and Cyber Unit," May 3, 2022, https://www.sec.gov/news/press-release/2022-78.

121 U.S. Securities and Exchange Commission, "Remarks Before the Aspen Security Forum," August 3, 2021, https://www.sec.gov/news/public-statement/gensler-aspen-security-forum-2021-08-03.

122 Olga Kharif and Joanna Ossinger, "Coinbase Gets Wells Notice From the SEC on Lend Product," Bloomberg, September 8, 2021, https://www.bloomberg.com/news/articles/2021-09-08/coinbase-gets-wells-notice-from-the-sec-on-its-lend-product.

123 Tweet by @brian_armstrong, September 7, 2021, https://twitter.com/brian_armstrong/status/1435439291715358721?lang=en.

124 Paul Grewal, "The SEC Has Told Us It Wants to Sue Us Over Lend. We Don't Know Why," *The Coinbase Blog*, September 7, 2021, https://blog.coinbase.com/the-sec-has-told-us-it-wants-to-sue-us-over-lend-we-have-no-idea-why-a3a1b6507009.

125 The White House, "FACT SHEET: President Biden to Sign Executive Order on Ensuring Responsible Development of Digital Assets," March 9, 2022, https://www.whitehouse.gov/briefing-room/statements-releases/2022/03/09/fact-sheet-president-biden-to-sign-executive-order-on-ensuring-responsible-innovation-in-digital-assets.

126 U.S. Congress, House of Representatives, *Infrastructure Investment and Jobs Act*, 117th Congress, 2021, H.Rep.3684, https://www.congress.gov/bill/117th-congress/house-bill/3684.

127 U.S. President, Legislation, "Infrastructure Investment and Jobs Act," H.R. 3684 (November 15, 2021), https://www.whitehouse.gov/briefing-room/legislation/2021/11/15/bill-signed-h-r-3684.

128 American Society of Civil Engineers, "ASCE's 2021 American Infrastructure Report Card GPA: C-," 2022, https://infrastructurereportcard.org.

129 U.S. Congress, Joint Committee On Taxation, *Estimated Revenue Effects of the Provisions in Division H of an Amendment in the Nature of a Substitute to H.R. 3684*, August 2, 2021, JCX-33-21, Title VI, https://www.jct.gov/CMSPages/GetFile.aspx?guid=f9c0b59d-de78-4173-993b-eb20b12ee5b8.

130 The White House, "FACT SHEET: The American Jobs Plan," March 31, 2022, https://www.whitehouse.gov/briefing-room/statements-releases/2021/03/31/fact-sheet-the-american-jobs-plan.

131 Kate Sullivan, "Biden Promises 'Once-In-A-Generation' Investment During Pitch for $2 Trillion Infrastructure and Climate Plan," CNN, March 31, 2021, https://edition.cnn.com/2021/03/31/politics/infrastructure-joe-biden-jobs-climate-plan/index.html.

132 The White House, "FACT SHEET: The American Jobs Plan," March 31, 2022, https://www.whitehouse.gov/briefing-room/statements-releases/2021/03/31/fact-sheet-the-american-jobs-plan.

133 Senator Rob Portman. "Rob's Rundown: Week of April 12 – April 16, 2021," April 16, 2021. https://www.portman.senate.gov/newsroom/robs-rundown/robs-rundown-week-april-12-april-16-2021.

134 Barbara Sprunt, "Biden's Bipartisan Infrastructure Package Fails A Test Vote In The Senate," NPR, July 21, 2021, https://www.npr.org/2021/07/21/1018278440/bipartisan-infrastructure-package-faces-1st-test-vote-in-senate.

135 The White House, "FACT SHEET: Historic Bipartisan Infrastructure Deal," July 28, 2021, https://www.whitehouse.gov/briefing-room/statements-releases/2021/07/28/fact-sheet-historic-bipartisan-infrastructure-deal.

136 Kristin Smith, Letter from Kristin Smith to Senators Charles Schumer and Mitch McConnell, July 30, 2021, Blockchain Association, https://theblockchainassociation.org/wp-content/uploads/2021/07/Blockchain-Association-Letter-to-Senate-Leadership-Re-Enhanced-Reporting-for-Brokers-and-Digital-Assets.pdf.

137 United States Senate Committee on Finance, "Wyden, Lummis, Toomey Amendment Would Clarify Digital Asset Reporting Requirements," August 4, 2021, https://www.finance.senate.gov/chairmans-news/wyden-lummis-toomey-amendment-would-clarify-digital-asset-reporting-requirements.

138 United States Senate, "U.S. Senate: Senate Floor Activity - Sunday, August 8, 2021," August, 8, 2021, https://www.senate.gov/legislative/LIS/floor_activity/2021/08_08_2021_Senate_Floor.htm.

139 James Pollard, "Texas Republicans Want to Make the State the Center of the Cryptocurrency Universe." *The Texas Tribune*, October 28, 2021, https://www.texastribune.org/2021/10/28/texas-republicans-blockchain-bitcoin/.

140 Turner Wright, "Infrastructure Bill Passes U.S. Senate—Without Clarification on Crypto," Cointelegraph, August 10, 2021, https://cointelegraph.com/news/infrastructure-bill-passes-us-senate-without-clarification-on-crypto.

141 United States Senate, *Senate Letter to Secretary Janet Yellen*, by Senators Rob Portman, Mark R. Warner, Mike Crapo, Kyrsten Sinema, Pat Toomey, and Cynthia Lummis, December 14, 2021.

142 The Law Library of Congress, *Regulation of Cryptocurrency Around the World: November 2021 Update*, November 2021, https://tile.loc.gov/storage-services/service/ll/llglrd/2021687419/2021687419.pdf.

143 Ibid.

144 Ibid.

145 Ibid.

146 "Deep in Rural China, Bitcoin Miners Are Packing Up," *The Economist*, July 10, 2021, https://www.economist.com/china/2021/07/10/deep-in-rural-china-bitcoin-miners-are-packing-up.

147 "China Bans Banks from Handling Bitcoin Trade," BBC News, December 5, 2013, https://www.bbc.com/news/technology-25233224.

148 The Central Bank of the Republic of Turkey, *Ödemelerde Kripto Varliklarin Kullanilmamasina*

Dair Yönetmelik, April 16, 2021, https://www.resmigazete.gov.tr/eskiler/2021/04/20210416-4.htm.

149 Jill Disis, Isil Sariyuce and Hanna Ziady, "Turkey's Lira Plunges after Erdogan Fires Central Bank Head," CNN, March 22, 2021, https://www.cnn.com/2021/03/22/economy/turkey-lira-erdogan-central-bank-intl-hnk/index.html.

150 Tito Sianipar, "Bitcoin Dilarang Otoritas Keuangan Indonesia, Ini Fakta-Faktanya," BBC News Indonesia, December 7, 2017, https://www.bbc.com/indonesia/indonesia-42265038.

151 Aastha Maheshwari, "Crypto Action Heats Up in Indonesia as Luno Mulls Market Entry, Tokyocrypto Weighs IPO," DealStreetAsia, June 1, 2021, https://www.dealstreetasia.com/stories/cryptocurrency-luno-tokocrypto-243052.

152 Otoritas Jasa Keuangan (@ ojkindonesia), "OJK Tegas Larang Lembaga Jasa Keuangan Fasilitasi Kripto," January 24, 2022, https://www.instagram.com/p/CZIgoP2PjI2/?hl=en.

153 Emmanuel Akinwotu, "Out of control and rising: why bitcoin has Nigeria's government in a panic," *The Guardian*, March 06, 2019, https://www.theguardian.com/technology/2021/jul/31/out-of-control-and-rising-why-bitcoin-has-nigerias-government-in-a-panic.

154 Raynor de Best, "Cryptocurrency Adoption Among Consumers—Statistics & Facts," Statista, March 29, 2021. https://www.statista.com/topics/7705/cryptocurrency-adoption-among-consumers.

155 Andrew S. Nevin and Omomia Omosomi, *Strength from Abroad: The Economic Power of Nigeria's Diaspora*, PwC Nigeria, 2019, https://www.pwc.com/ng/en/pdf/the-economic-power-of-nigerias-diaspora.pdf.

156 The World Bank, "Defying Predictions, Remittance Flows Remain Strong During COVID-19 Crisis," World Bank press release, May 12, 2021, https://www.worldbank.org/en/news/press-release/2021/05/12/defying-predictions-remittance-flows-remain-strong-during-covid-19-crisis.

157 Bello Hassan and Muso I. Jimoh, Letter from Bello Hassan and Muso I. Jimoh to all deposit money banks, non-financial institutions and other financial institutions, February 5, 2021, Central Bank of Nigeria, https://www.cbn.gov.ng/out/2021/ccd/letter%20on%20crypto.pdf.

158 Board of Governors of the U.S. Federal Reserve System, "What Is a Central Bank Digital Currency?", January 20, 2022, https://www.federalreserve.gov/faqs/what-is-a-central-bank-digital-currency.htm.

159 Central Bank Digital Currency (CBDC) Tracker, "Central Bank Digital Currency (CBDC) Tracker," June 19, 2022, https://cbdctracker.org.

160 U.S. Federal Reserve, *Money and Payments: The U.S. Dollar in the Age of Digital Transformation*, January 2022, https://www.federalreserve.gov/publications/files/money-and-payments-20220120.pdf.

161 Iori Kawate, "China's Digital Yuan Pilot Tally Reaches $5.3bn in Six Months," Nikkei Asia, July 17, 2021, https://asia.nikkei.com/Economy/China-s-digital-yuan-pilot-tally-reaches-5.3bn-in-six-months.

162 People's Bank of China, "Progress of Research & Development of E-CNY in China," July 2021, http://www.pbc.gov.cn/en/3688110/3688172/4157443/4293696/2021071614584691871.pdf.

163 Jung Min-kyung, "BOK Says Mock Issuance, Distribution of Digital Currency Successful," *The Korea Herald*, January 24, 2022, https://www.koreaherald.com/view.php?ud=20220124000761.

164 Ibid.

165 "China Bans Financial Companies From Bitcoin Transactions," Bloomberg, December 5, 2013, https://www.bloomberg.com/news/articles/2013-12-05/china-s-pboc-bans-financial-companies-from-bitcoin-transactions.

166 "Cryptocurrency Mining Declared Obsolete by China's State Economic Planner," China Banking News, January 13, 2022, https://www.chinabankingnews.com/2022/01/13/cryptocurrency-mining-declared-obsolete-by-chinas-state-economic-planner.

167 Xie Yu, "China to Stamp Out Cryptocurrency Trading Completely with Ban on Foreign Platforms," *South China Morning Post*, February 5, 2018, https://www.scmp.com/business/banking-finance/article/2132009/china-stamp-out-cryptocurrency-trading-completely-ban.

168 Wolfie Zhao, "China's Economic Planning Body Labels Bitcoin Mining an 'Undesirable' Industry," CoinDesk, April 9, 2019, https://www.coindesk.com/markets/2019/04/09/chinas-economic-planning-body-labels-bitcoin-mining-an-undesirable-industry.

169 Library of Congress, "China: National Development and Reform Commission Issues Notice Restricting Cryptocurrency Mining," September 24, 2021, https://www.loc.gov/item/global-legal-monitor/2022-02-08/china-national-development-and-reform-commission-issues-notice-restricting-cryptocurrency-mining/.

170 Ibid.

171 Library of Congress, "China: Central Bank Issues New Regulatory Document on Cryptocurrency Trading," September 24, 2021, https://www.loc.gov/item/global-legal-monitor/2021-10-13/china-central-bank-issues-new-regulatory-document-on-cryptocurrency-trading/.

172 Laura He, "Bitcoin falls as China takes aim once again at 'extremely harmful' crypto mining," CNN, November 16, 2021, https://www.cnn.com/2021/11/16/investing/bitcoin-china-crypto-mining-crackdown-intl-hnk/index.html.

173 "12 Facts on China's Economic History." *The Globalist*, November 10, 2014, https://www.theglobalist.com/12-facts-on-chinas-economic-history/.

174 "China's Great Leap Forward, 1958–1961," Wilson Center, n.d., https://digitalarchive.wilsoncenter.org/collection/210/china-s-great-leap-forward-1958-1961.

175 Clayton D. Brown, "China's Great Leap Forward," *Education About Asia*, 17, no. 3 (2012), https://www.asianstudies.org/publications/eaa/archives/chinas-great-leap-forward/.

176 Michael McSweeney, "Bitcoin Miners Generated More than $15 Billion in Revenue During 2021," The Block, December 23, 2021, https://www.theblockcrypto.com/linked/128475/bitcoin-mining-2021-revenue.

177 Cao LI, "China, a Major Bitcoin Source, Considers Moving Against It," *The New York Times*, April 9, 2019, https://www.nytimes.com/2019/04/09/business/bitcoin-china-ban.html.

178 Joanna Ossinger, J. and Zheping Huang, "Chinese Regulators Are Serious About Crypto Ban This Time," Bloomberg, September 26, 2021, https://www.bloomberg.com/news/articles/2021-09-26/chinese-regulators-are-serious-about-banning-crypto-this-time.

179 Coco Feng, "BTCChina, The Country's First Bitcoin Exchange, Gives Up on the Cryptocurrency Amid Beijing's Tightening Crackdown," *South China Morning Post*, June 24, 2021, https://www.scmp.com/tech/policy/article/3138618/btcchina-countrys-first-bitcoin-exchange-gives-cryptocurrency-amid.

180 Laura He, "Bitcoin Falls as China Takes Aim Once Again at 'Extremely Harmful' Crypto Mining," CNN, November 16, 2021, https://www.cnn.com/2021/11/16/investing/bitcoin-china-crypto-mining-crackdown-intl-hnk/index.html.

181 Charles Bovaird, "Bitcoin Prices Reached An All-Time High Above $63,000—What's Next?", Forbes, April 13, 2021, https://www.forbes.com/sites/cbovaird/2021/04/13/bitcoin-prices-reached-an-all-time-high-above-63000-whats-next/.

182 "Deep in Rural China, Bitcoin Miners Are Packing Up," *The Economist*, July 10, 2021, https://www.economist.com/china/2021/07/10/deep-in-rural-china-bitcoin-miners-are-packing-up.

183 "Bitcoin USD (BTC-USD) Price History & Historical Data," Yahoo Finance, Accessed August 12, 2022, https://finance.yahoo.com/quote/BTC-USD/history/. 135???

184 "The Great Firewall of China," Bloomberg, November 5, 2018, https://www.bloomberg.com/quicktake/great-firewall-of-china.

185 Number of Journalists Behind Bars Reaches Global High," Committee to Protect Journalists, December 9, 2021, https://cpj.org/reports/2021/12/number-of-journalists-behind-bars-reaches-global-high/.

186 Saheli Roy Choudhury, "The Yuan Hit an 11-Year Low This Week. Here's a Look at How China Controls Its Currency," August 28, 2019, https://www.cnbc.com/2019/08/28/china-economy-how-pboc-controls-the-yuan-rmb-amid-trade-war.html.

187 "China's Capital Controls: Here to Stay?," Central Banking, July 30, 2021, https://www.centralbanking.com/node/7860946.

188 Shunsuke Tabeta and Iori Kawate, "Xi Raises Prospect of Doubling China's GDP by 2035," Nikkei Asia, November 4, 2020, https://asia.nikkei.com/Economy/Xi-raises-prospect-of-doubling-China-s-GDP-by-2035.

189 Congressional Research Service, *The U.S. Dollar as the World's Dominant Reserve Currency*, December 18, 2020, https://crsreports.congress.gov/product/details?prodcode=IF11707.

190 "IMF Adds Chinese Renminbi to Special Drawing Rights Basket," International Monetary Fund, September 30, 2016, https://www.imf.org/en/News/Articles/2016/09/29/AM16-NA093016IMF-Adds-Chinese-Renminbi-to-Special-Drawing-Rights-Basket.

191 "Office of Foreign Assets Control—Sanctions Programs and Information," U.S. Department of the Treasury, n.d., https://home.treasury.gov/policy-issues/office-of-foreign-assets-control-sanctions-programs-and-information.

192 Yen Nee Lee, "Faced With a Power Crisis, China May Have 'Little Choice' But to Ramp Up Coal Consumption," CNBC, October 17, 2021, https://www.cnbc.com/2021/10/18/power-crunch-china-has-little-choice-but-increase-coal-use-analysts-say.html.

193 U.S. Embassy in Georgia, "China's Air Pollution Harms Its Citizens and the World," November 24, 2021, https://ge.usembassy.gov/chinas-air-pollution-harms-its-citizens-and-the-world/.

194 IEA Executive Director and China's Special Envoy on Climate Change Discuss Global Efforts to Reach Net-Zero Emissions," IEA, June 25, 2021, https://www.iea.org/news/iea-executive-director-and-china-s-special-envoy-on-climate-change-discuss-global-efforts-to-reach-net-zero-emissions.

195 "Global Trends in Renewable Energy Investment 2019", UN Environment Programme, September 11, 2019.

196 "Profiling the Top Five Countries with the Highest Wind Energy Capacity," NS Energy, March 30, 2021, https://www.nsenergybusiness.com/features/top-countries-wind-energy-capacity/.

197 Ferdinand Bada, "The Largest Hydroelectric Power Stations in China," WorldAtlas, July 12, 2018, https://www.worldatlas.com/articles/the-largest-hydroelectric-power-stations-in-china.html.

198 "China Is the World's Largest Producer of Hydroelectricity," Hydropower. Org, 2022, https://www.hydropower.org/country-profiles/china.

199 Samuel Shen & Andrew Galbraith, "China's Ban Forces Some Bitcoin Miners to Flee Overseas, Others Sell Out," Reuters, June 25, 2021. https://www.reuters.com/technology/chinas-ban-forces-some-bitcoin-miners-flee-overseas-others-sell-out-2021-06-25/.

200 Frankfurt School, United Nations Environment Programme Collaborative Center for Climate & Sustainable Energy Finance, *Global Trends in Renewable Energy Investment 2019*, 2019, https://www.fs-unep-centre.org/wp-content/uploads/2019/11/GTR_2019.pdf.

201 Lisa Minton, "Texas' Electricity Resources: Where Power Comes From—and How It Gets to You," August 2020, https://comptroller.texas.gov/economy/fiscal-notes/2020/august/ercot.php.

202 Alvaro Trigueros-Argüello and Marjorie Chorro de Trigueros, "Bitcoin as Legal Tender in El Salvador: The First Fifty Days," Georgetown Journal of International Affairs, November 30, 2021, https://gjia.georgetown.edu/2021/11/30/bitcoin-as-legal-tender-in-el-salvador-the-first-fifty-days/.

203 "El Salvador," MapFlight, n.d., https://mapfight.xyz/map/sv/.

204 "El Salvador," The Center for Justice & Accountability, n.d., https://cja.org/where-we-work/el-salvador/.

205 United States Institute of Peace, *Truth Commission: El Salvador*, July 1, 1992. https://www.usip.org/publications/1992/07/truth-commission-el-salvador.

206 "Hurricane Mitch," History, Updated: November 11, 2019, https://www.history.com/topics/natural-disasters-and-environment/hurricane-mitch.

207 "Today in Earthquake History," USGS, n.d., https://earthquake.usgs.gov/learn/today/index.php?month=1&day=13&submit=View+Date.

208 "El Salvador," CIA: The World Factbook, August 9, 2022, https://www.cia.gov/the-world-factbook/countries/el-salvador/#introduction.

209 "P3 Legislation in El Salvador: An Aggressive Reassertion of Neoliberal Economics?," Council on Hemispheric Affairs, August 7, 2013, https://www.

coha.org/p3-legislation-in-el-salvador-an-aggressive-reassertion-of-neoliberal-economics/.

210 Elaine Freedman, "Public-Private Partnerships: Another Disguise for Privatization," Envio Digital, June 2012, https://www.envio.org.ni/articulo/4542.

211 "The Strange Political Path of Nayib Bukele, El Salvador's New President," KXLY, February 11, 2019, https://www.kxly.com/the-strange-political-path-of-nayib-bukele-el-salvadors-new-president/.

212 Ibid.

213 "Bitcoin Beach," Bitcoin Beach, n.d., https://www.bitcoinbeach.com/.

214 La Asamblea Legislativa de la República de El Salvador, *Ley de Creación del Fideicomiso Bitcoin,* June 9, 2021, https://www.transparencia.gob.sv/system/documents/documents/000/450/425/original/Ley_de_Creaci%C3%B3n_del_Fideicomiso_Bitcoin.pdf?1631049624.

215 Friedemann Brenneis, "Why El Salvador Is Banking on Bitcoin," The Red Bulletin, January 4, 2022, https://www.redbull.com/za-en/theredbulletin/el-salvador-bitcoin-currency#:~:text=Around%2070%20percent%20of%20El,national%20and%20international%20payment%20transactions.

216 "History," International Monetary Fund, n.d., https://www.imf.org/external/about/history.htm.

217 Eric Martin, "IMF Urges El Salvador to Strip Bitcoin's Legal Tender Status," Bloomberg, January 25, 2022, https://www.bloomberg.com/news/articles/2022-01-25/imf-board-urges-el-salvador-to-ditch-bitcoin-as-legal-tender.

218 Antonia Noori Farzan, "World Bank Declines to Help El Salvador Adopt Bitcoin, Citing Environmental and Transparency Concerns," *Washington Post*, June 17, 2021, https://www.washingtonpost.com/world/2021/06/17/world-bank-bitcoin-el-salvador/.

219 Andres Engler, "Panama to Present Crypto-Related Bill in July," CoinDesk, June 16, 2021, https://www.coindesk.com/markets/2021/06/16/panama-to-present-crypto-related-bill-in-july/.

220 Tweet by @carlitosrejala, June 6, 2021, https://twitter.com/carlitosrejala/status/1401712725886132224.

221 Tweet by @eduardomurat, June 8, 2021, https://twitter.com/eduardomurat/status/1402257568580390923.

222 Arjun Kharpal, "El Salvador Plans to Create a 'Bitcoin City' and Raise $1 Billion via a 'Bitcoin Bond,'" CNBC, November 22, 2021, https://www.cnbc.com/2021/11/22/el-salvador-plans-bitcoin-city-raise-1-billion-via-bitcoin-bond.html.

223 Bhushan Akolkar. "El Salvador Announces Bitcoin City, Plans to Issue $1 Billion in Tokenized Bitcoin Bonds." Binance, November 21, 2021. https://www.binance.com/en/news/top/6500902.

224 International Monetary Fund, *El Salvador: Staff Concluding Statement of the 2021 Article IV Mission*, November 22, 2021, https://www.imf.org/en/News/Articles/2021/11/22/mcs-el-salvador-staff-concluding-statement-of-the-2021-article-iv-mission.

225 International Monetary Fund, "Executive Board Concludes 2021 Article IV Consultation with El Salvador," January 25, 2022, https://www.imf.org/en/News/Articles/2022/01/25/pr2213-el-salvador-imf-executive-board-concludes-2021-article-iv-consultation.

226 Tweet by @nayibbukele, January 25, 2022, https://twitter.com/nayibbukele/status/1486162932224479235.

227 "S&P Global Ratings Revised Outlook on El Salvador to Negative and Affirmed at "B-" (Local Currency LT) Credit Rating," October 25, 2021, https://cbonds.com/news/1481695/.

228 "Rating Action: Moody's Downgrades El Salvador's Rating to Caa1, Maintains Negative Outlook," Moody's Investor Services, July 30, 2021, https://www.moodys.com/research/Moodys-downgrades-El-Salvadors-rating-to-Caa1-maintains-negative-outlook--PR_450956.

229 "Bitcoin Implementation a Credit Negative for El Salvador Insurers," FitchRatings, August 16, 2021, https://www.fitchratings.com/research/insurance/bitcoin-implementation-credit-negative-for-el-salvador-insurers-16-08-2021.

230 Unicef, *UNICEF El Salvador Country Office Annual Report 2021*, 2021, https://www.unicef.org/media/116261/file/El-Salvador-2021-COAR.pdf.

231 Ibid.

232 Ibid.

233 Ibid.

234 "El Salvador Events Of 2020," Human Rights Watch, 2022, https://www.hrw.org/world-report/2021/country-chapters/el-salvador.

235 Ibid.

236 "El Salvador," World Prison Brief, 2021, https://www.prisonstudies.org/country/el-salvador.

237 Tweet by @PNCSV, March 26, 2022, https://twitter.com/pncsv/status/1507595993432342528?s=21&t=9NUNOGuCy66mOopupfgAlQ.

238 Tweet by @PNCSV, March 27, 2022, https://twitter.com/pncsv/status/1507972267304787968?s=12&t=5QvGSTzIeSpnz9-zLTU1Xw.

239 Tweet by @PNCSV, March 28, 2022, https://twitter.com/PNCSV/
     status/1508319971419566080.

240 U.S. Embassy in El Salvador, *Notification to U.S. Citizens Resident In or
     Traveling to El Salvador*, March 29, 2022, https://sv.usembassy.
     gov/u-s-embassy-san-salvador-el-salvador-march-29-2022/.

241 Tweet by @PNCSV, March 26, 2022, https://twitter.com/PNCSV/
     status/1526084389712568320.

242 "El Salvador: Response to the Rise in Gang Killings 'Cruel and Inhuman',"
     UN News, April 5, 2022, https://news.un.org/en/story/2022/04/1115562.

243 U.S. Department of the Treasury, "Treasury Issues Sanctions on International
     Anti-Corruption Day," December 9, 2021, https://home.treasury.gov/news/
     press-releases/jy0523.

244 Ibid.

245 Ibid.

246 David Bernal, "Sanction to Carolina Recinos would affect the Chivo Wallet",
     La Presnsa Grafica, December 10, 2021. Sanction to Carolina Recinos would
     affect the Chivo Wallet (laprensagrafica.com).

247 "Personal Remittances, Received (% of GDP)—El Salvador," The World
     Bank, 2020, https://data.worldbank.org/indicator/BX.TRF.PWKR.DT.GD.
     ZS?locations=SV.

248 Jaime Quintanilla, "Miembros de Nuevas Ideas al Frente de Empresa que
     Administra Chivo Wallet," El Economista, September 9, 2021. https://www.
     eleconomista.net/economia/El-Salvador-miembros-de-Nuevas-Ideas-al-
     frente-de-empresa-que-administra-Chivo-Wallet-20210909-0001.html.

249 Jorge Beltran Luna, "Empresa Chivo was Created with Public Funds from
     CEL," September 8, 2021, https://www.elsalvador.com/noticias/nacional/
     empresa-chivo-creada-fondos-publicos-cel-bitcoin/877091/2021/.

250 Nate DiCamillo, "State Lawmaker Explains Wyoming's Newly Passed
     DAO LLC Law," CoinDesk, April 22, 2021, https://www.coindesk.com/
     policy/2021/04/22/state-lawmaker-explains-wyomings-newly-passed-dao-llc-
     law/.

251 Anessa Allen Santos, *Wyoming Blockchain Legislation Summary Review for
     Years 2018–2019*, Business Law Section of the Florida Bar, 2019, http://
     www.flabizlaw.org/files/Wyoming%20Blockchain%20Legislation%20
     Summary%20Review.pdf.

252 Andrew Glass, "Wyoming Becomes 44th State July 10, 1890," Politico, July
     10, 2007, https://www.politico.com/story/2007/07/wyoming-becomes-44th-
     state-july-10-1890-004845.

253 "Wyoming Population 2022," World Population View, 2022, https://worldpopulationreview.com/states/wyoming-population.

254 "Wyoming Information," Cody Yellowstone, 2022, https://www.codyyellowstone.org/plan/wyoming-information/.

255 "Taxes in Wyoming," Tax Foundation, 2022, https://taxfoundation.org/state/wyoming/#:~:text=Wyoming%20Tax%20Rates%2C%20Collections%2C%20and%20Burdens&text=Wyoming%20does%20not%20have%20an,tax%20rate%20of%205.22%20percent.

256 "Wyoming," Netstate, January 14, 2018, https://www.netstate.com/states/intro/wy_intro.htm#:~:text=Wyoming%20was%20nicknamed%20the%20%22Equality,juries%20and%20hold%20public%20office.

257 "Wyoming Legislators Write the First State Constitution to Grant Women the Vote," History, September 28, 2021, https://www.history.com/this-day-in-history/wyoming-legislators-write-the-first-state-constitution-to-grant-women-the-vote.

258 "Nellie Tayloe Ross Governor of Wyoming, United States," Britannica, June 28, 2022, https://www.britannica.com/biography/Nellie-Tayloe-Ross.

259 "Yellowstone Park Established," History, February 28, 2022, https://www.history.com/this-day-in-history/yellowstone-park-established; "Equal Rights," State Symbols U.S.A., n.d., https://statesymbolsusa.org/wyoming/state-motto/wyoming-state-motto.

260 "The Complete History of the LLC," Wyoming LLC, 2022, https://www.wyomingllcs.com/history-of-the-llc/.

261 "Welcome to the Shoshone National Forest," USDA Forest Service, n.d., https://www.fs.usda.gov/shoshone.

262 North Antelope Rochelle Coal Mine, Wyoming," NASA, n.d., https://earthobservatory.nasa.gov/images/5915/north-antelope-rochelle-coal-mine-wyoming.

263 Carmen, "World's Ten Largest Coal Mines in 2020," Mining Technology, September 7, 2021, https://www.mining-technology.com/marketdata/ten-largest-coals-mines-2020/.

264 Amy Tikkanen, "J.C. Penney," Britannica, n.d., https://www.britannica.com/topic/JC-Penney-Corporation-Inc.

265 Caitlin Long, "About Caitlin," 2019, https://caitlin-long.com/about-caitlin/.

266 Caitlin Long, "What Do Wyoming's 13 New Blockchain Laws Mean?," Forbes, March 4, 2019, https://www.forbes.com/sites/caitlinlong/2019/03/04/what-do-wyomings-new-blockchain-laws-mean/?sh=5772135a5fde.

267 "Part 200 Virtual Currencies," Thomson Reuters Westlaw, n.d. https://govt.

westlaw.com/nycrr/Browse/Home/NewYork/NewYorkCodesRulesandRegulations?guid=I7444ce80169611e594630000845b8d3e&originationContext=documenttoc&transitionType=Default&contextData=(sc.Default)

268 Yessi Bello Perez, "The Real Cost of Applying for a New York BitLicense," CoinDesk, September 11, 2021, https://www.coindesk.com/markets/2015/08/13/the-real-cost-of-applying-for-a-new-york-bitlicense/.

269 Department of Financial Services. 2022. *Industry Letter - June 24, 2020: Virtual Currency Guidance - Notice of Virtual Currency Business Activity License Application Procedures.* [online] Available at: <https://www.dfs.ny.gov/industry_guidance/industry_letters/il20200624_notice_vc_busact_lic_app_procedure> [Accessed 8 October 2022].

270 Ibid.

271 New York State Department of Financial Services, List of Regulated Entities In Virtual Currencies, https://www.dfs.ny.gov/virtual_currency_businesses#regulated_entities

272 Caitlin Long, "What Do Wyoming's 13 New Blockchain Laws Mean?," March 4, 2019, https://caitlin-long.com/what-do-wyomings-13-new-blockchain-laws-mean/.

273 Ibid.

274 State of Wyoming Legislature, *HB0019 - Wyoming Money Transmitter Act—Virtual Currency Exemption*, 2018, https://www.wyoleg.gov/Legislation/2018/HB0019.

275 NMLS, *WY Money Transmitter License New Application Checklist (Company)*, April 20, 2022, https://mortgage.nationwidelicensingsystem.org/slr/PublishedStateDocuments/WY-Money-Transmitter-Company-New-App-Checklist-Final.pdf.

276 State of Wyoming Legislature, 2022 Select Committee on Blockchain, Financial Technology and Digital Innovation Technology, https://wyoleg.gov/Committees/2022/S19)

277 State of Wyoming Legislature, *Special Purpose Depository Institutions*, HB0074, 2019, https://www.wyoleg.gov/Legislation/2019/hb0074.

278 Wyoming Division of Banking, *Special Purpose Depository Institutions*, n.d., https://wyomingbankingdivision.wyo.gov/banks-and-trust-companies/special-purpose-depository-institutions.

279 Jeff Benson, "Avanti Becomes Second Crypto Bank in US," Decrypt, October 28, 2020, https://decrypt.co/46561/avanti-becomes-second-crypto-bank-us.

280 "Breaking: SEC Investigating Binance Coin (BNB)," Crypto News, June 6, 2022, https://cryptonews.net/news/regulation/7798075/.

281 Stan Higgins, "$257 Million: Filecoin Breaks All-Time Record for ICO Funding," CoinDesk, September 7, 2017, https://www.coindesk.com/markets/2017/09/07/257-million-filecoin-breaks-all-time-record-for-ico-funding/.

282 Evelyn Cheng, "JPMorgan Chase, Bank of America & Citi Bar People from Buying Bitcoin with a Credit Card," CNBC, February 3, 2018, https://www.cnbc.com/2018/02/02/jpmorgan-chase-bank-of-america-bar-bitcoin-buys-with-a-credit-card.html.

283 u/EliToohey, "Just got a call from my bank demanding I tell them why I purchased Bitcoin or they'll close my account," 2017, https://www.reddit.com/r/Bitcoin/comments/75tm44/just_got_a_call_from_my_bank_demanding_i_tell/.

284 David Fry, "Why Bitcoin Access Has Been Shut Down in Hawaii," Honolulu Civil Beat, March 14, 2017, https://www.civilbeat.org/2017/03/why-bitcoin-access-has-been-shut-down-in-hawaii/.

285 Tim Enneking, "Bank of America Is Closing My Three-Year-Old's Account Over Crypto," CoinDesk, April 28, 2018, https://www.coindesk.com/markets/2018/04/28/bank-of-america-is-closing-my-three-year-olds-account-over-crypto/.

286 Kate Rooney and Evelyn Chang, "Meet The Small Community Lender That's Become the Go-To Banker of the Cryptocurrency World," May 31, 2018, https://www.cnbc.com/2018/05/31/meet-silvergates-alan-lane-whos-bankrolling-cryptocurrency-exchanges.html.

287 U.S. Federal Reserve, *Federal Reserve Announces Members of Its Community Depository Institutions Advisory Council for 2022*, January 28, 2022, https://www.federalreserve.gov/newsevents/pressreleases/other20220128a.htm.

288 u/dan_from_san_diego, "Chase Is Closing My Account Due to Bitcoin Purchases. Nice.," 2016, https://www.reddit.com/r/Bitcoin/comments/5mzwla/chase_is_closing_my_account_due_to_bitcoin/.

289 Community Depository Institutions Advisory Council and the Board of Governors, *Record of Meeting*, November 18, 2021, https://www.federalreserve.gov/aboutthefed/files/CDIAC-meeting-20211118.pdf.

290 U.S. Senate Committee on Banking, Housing, and Urban Affairs, *Nomination Hearing*, January 11, 2022, https://www.banking.senate.gov/hearings/01/04/2022/nomination-hearing.

291 Sam Wouters, "Top Banks Investing in Crypto & Blockchain Companies," Blockdata, May 9, 2021, https://www.blockdata.tech/blog/general/banks-investing-blockchain-companies.

292 Nikhilesh De, "Caitlin Long's Wyoming Crypto Bank Takes a Step Toward Fed Membership," CoinDesk, February 10, 2022, https://www.coindesk.

com/policy/2022/02/10/caitlin-longs-crypto-bank-takes-a-step-toward-fed-
membership/.

293 "ConstitutionDAO," 2021, https://www.constitutiondao.com/.

294 Nathan Reiff, "Decentralized Autonomous Organization (DAO),"
     Investopedia, July 11, 2022, https://www.investopedia.com/tech/what-dao/.

295 State of Wyoming Legislature, *Decentralized Autonomous Organizations*,
     SF0038, 2021, https://www.wyoleg.gov/Legislation/2021/SF0038.

296 State of Wyoming Legislature, *Wyoming Chancery Court*, SF0104, 2019,
     https://www.wyoleg.gov/Legislation/2019/SF0104.

297 Wyoming Secretary of State's Office, "DAO's In Wyoming", Business@wyo.
     gov, 2022.

298 State of Wyoming Legislature, Wyoming Stable Token Act, SF0106, 2022,
     https://wyoleg.gov/Legislation/2022/SF0106.